W0225644

In Connection with

Biophysics of **Structure** and **Mechanism**

Radiation and Environmental Biophysics

Annual Meeting of the
Deutsche Gesellschaft für Biophysik

Konstanz, October 1979

Abstracts of Poster Presentations

Edited by G. Adam and G. Stark

Springer-Verlag Berlin Heidelberg GmbH 1979

Editors:
Professor Dr. Gerold Adam
Fakultät für Biologie der Universität Konstanz, Universitätsstraße 10, D-7750 Konstanz

Professor Dr. Günther Stark
Fakultät für Biologie der Universität Konstanz, Universitätsstraße 10, D-7750 Konstanz

ISBN 978-3-540-09684-9   ISBN 978-3-642-51881-2 (eBook)
DOI 10.1007/978-3-642-51881-2

CONTENTS

Section A: Molecular Biophysics   1

A1   K.-D. Kohl and W. Sperling   2
7-cis retinal, an artificial chromophore of bacteriorhodopsin

A2   G. Drikos, H. Rüppel and W. Sperling   3
On the determination of the absorption spectra of some
11-cis-retinal conformeres by polarisation spectroscopy
using extremely thin crystal plates

A3   J. Tretzel and F.W. Schneider   4
Resonance CARS spectroscopy of bacteriorhodopsin

A4   W. Mäntele, F. Siebert and W. Kreutz   5
Kinetic infrared spectroscopy on the photocycle of bacteriorhodopsin.

A5   F. Siebert, W. Mäntele and W. Kreutz   6
Kinetic infrared investigation on the photodissociation
of sperm whale CO-myoglobin

A6   P. Grunwald, W. Gunsser and E. Schober   7
Some theoretical and practical aspects of enzyme immobilisation at the example urease/activated aluminiumhydro-
xide

A7   P. Grunwald and W. Gunsser   8
Kinetic of urease-catalysed urea hydrolysis in the presence of alkaline earth halides

A8   D. Schallreuter and E. Neumann   9
Ion atmosphere relaxation of simple electrolytes and polyionic biopolymers in electric fields

A9   C.-R. Rabl   10
High-resolution temperature-jump measurements in biochemistry

A10   H. Klump, Th. Ackermann, V. Gramlich, Th. Knäble,
E.D. Schmid, H. Seliger and J. Stulz   11
Demonstration of G·U wobble base pairs by Raman and IR
spectroscopy

A11   W. Köhnlein, R.S. Lewis and G. Jung   12
A new model for the activation and inactivation of neocarzinostatin, an antitumor protein

A12   G. Jung, R.S. Lewis and W. Köhnlein   13
Generation of pre-neocarzinostatin and its antagonistic
effect on neocarzinostatin-induced DNA strand scission

A13   F. Zimmermann and W. Köhnlein   14
The production of strand breaks by the interaction of
bleomycin with supercoiled col El DNA: An unexpected
concentration dependence

A14   M. Christahl, K. Gersonde, A. Raap and C. Appleby   15
Electron spin resonance study of cobalt leghaemoglobin

A15   H. Sapper, W. Gohl and W. Lohmann   16
Interactions of ATP phosphate groups with amines and
divalent metal ions: ^{31}P-NMR studies

A16   W. Wallbott, H. Neubacher and W. Lohmann   17
Investigations on the interaction of NADH with Cu(II):
ESR and optical absorption spectroscopy

A17   P. Zaplatynski, H. Neubacher and W. Lohmann   18
Experiments on the reaction kinetics between "platinum blue" compounds and several biomolecules

A18   H. Neubacher, P. Zaplatynski and W. Lohmann   19
Investigations on the molecular interaction of the an-
ticarcinogenic compound cis-dichlorodiammineplatinum
(II): The formation of colored platinum complexes

A19   A. Bahr, V. Penka and W. Lohmann   20
Modification of the radiation effect on nucleobases
and nucleosides by nitroimidazole-derivatives and
ascorbic acid

A20   W. Karsten and A. Müller   21
Radikalerzeugung in Einkristallen decarboxylierter
aromatischer Aminosäuren

Section B: Cell- and Membrane Research   22

B1   O. Albrecht, H. Gruler and E. Sackmann   23
Polymorphism of pure and mixed phospholipid monolayers

B2   E. Jürgens and E. Sackmann   24
Spectroscopic investigations of macroscopically ordered
lecithin - water - multibilayers

B3   W. Knoll, E. Sackmann and H.B. Stuhrmann   25
Neutron small angle scattering on model membranes

B4   A. Blume, B. Gruenewald and F. Watanabe   26
Kinetic investigations of the phase transition of lecithin bilayers

B5   A. Pleyer-Weber, H. Sapper, W. Strobelt and W. Lohmann   27
The interaction between L-ascorbic acid, α-tocopherol,
and phospholipid model membranes

B6   H. Vogel   28
The interaction of synthetic phospholipids with melittin
as a model to study lipid-protein interactions

B7   B. Tümmler, U. Hermann, G. Maaß and F. Vögtle   29
Einfluß künstlicher Ionophore auf die Struktur von Modell-
membranen

B8   H. Akutsu, H. Satake, R.M. Franklin and J. Seelig   30
^{31}-P-NMR studies on the lipid containing bacteriophage PM2

B9   J.C.W. Shepherd and G. Büldt   31
The head group mobility in phosphatidylcholine membranes
in the presence of cholesterol as seen by a dielectric investigation

B10   G. Büldt, H.U. Gally, A. Seelig, J. Seelig and G. Zaccai   32
A time averaged picture of the conformation and segmental
disorder of phosphatidylcholine in bilayers investigated
by neutron diffraction

B11   M.P. Heyn   33
       Determination of lipid order parameters from fluorescence
       depolarization experiments

B12   M.P. Heyn, N.A. Dencher and R.J. Cherry   34
       Lipid-protein interactions in bacteriorhodopsin-phospha-
       tidylcholine vesicles

B13   N.A. Dencher and M.P. Heyn   35
       Aggregated and monomeric bacteriorhodopsin pumps protons

B14   S. Stankowski and B. Gruenewald   36
       On the interpretation of lipid membrane cooperativity data

B15   J.-P. Meraldi, J. Schlitter and J. Seelig   37
       Statistical description of phospholipid bilayers with
       application to deuterium NMR

B16   F. Jähnig   38
       Structural order of lipids and proteins in membranes.
       New evaluation of fluorescence anisotropy data

B17   R. Lawaczeck   39
       New measurements of the water-permeability across lipid
       bilayer vesicles, the effect of cholesterol

B18   A. Haase and P. Fromherz   40
       Investigations of lipoid pH indicators as probes for
       electrostatic potential of micelles and vesicles

B19   A. Hermetter and F. Paltauf   41
       Influence of cholesterol on permeability and structural
       properties of artificial membranes containing enantio-
       meric phosphatidylcholines and ether analogues

B20   P. Jauch, A. Hermetter, F. Paltauf and R. Benz   42
       Surface potential studies of monolayers from different
       sterols and various phospholipid/sterol mixtures

B21   R. Schindler and R. Benz   43
       Influence of surface potential on ion transport through
       lipid bilayer membranes

B22   G. Ehmann and R. Benz   44
       Influence of local anesthetics on ion transport through
       lipid bilayer membranes

B23   G. Stark   45
       Negative hydrophobic ions as carriers for positive hy-
       drophobic ions

B24   W. Brock, P.C. Jordan and G. Stark   46
       Laser-temperature-jump method for the determination of
       the rate of desorption of hydrophobic ions in lipid mem-
       branes

B25   R. Junges and H.-A. Kolb   47
       Voltage dependent noise current of lipid bilayer mem-
       branes generated by hydrophobic ions

B26   H.-A. Kolb and E. Frehland   48
       Noise analysis of carrier-mediated ion transport under
       nonequilibrium conditions

B27   H.-A. Kolb and D. Woermann   49
       1/f noise in track-etched mica membranes

B28   E. Frehland   50
       Current fluctuations in biological transport systems far
       from equilibrium

B29   W. Stephan and E. Frehland   51
       Ion transport through pores: oscillatory phenomena at
       transient states

B30   H.-H. Kohler   52
       Why single file pores?

B31   E. Bamberg and K. Janko   53
       Properties of ionic channels made from derivatives of
       gramicidin A

B32   W. Hanke and G. Boheim   54
       State dependent blocking effects on the alamethicin pore
       by divalent cations

B33   H. Brückner, G. Jung, W. Hanke and G. Boheim   55
       Structural requirements for membrane modifying activity
       in alamethicin-type antibiotics

B34   S. Oberschär and G. Boheim   56
       Investigations on monolayers of alamethicin-type antibio-
       tics at an air/water interphase

B35   H. Craubner   57
       Transport of matter through membranes. The influence of
       protein on the kinetics of the free dialysis of detergents

B36   H. Schindler   58
       Vesicle spread membranes, a novel way for membrane re-
       constitution

B37   H. Schindler   59
       Matrix protein/ATPase proteolipid/acetylcholine receptor
       in vesicle spread membranes

B38   A. Schneider and E. Bamberg   60
       Formation and characterization of bilayer membranes from
       vesicle-spread monolayers

B39   A. Fahr and E. Bamberg   61
       Studies on photocurrent kinetics of purple membrane on
       black lipid membranes

B40   H.-J. Dorst, G. Pappert, D. Schubert and R. Benz   62
       Association equilibria between oligomers of band 3-pro-
       tein from human erythrocyte membranes: Time needed for
       relaxation towards equilibrium

B41   R. Benz and T. Nakae   63
       Properties of large ion-permeable pores formed by porins
       from Salmonella typhimurium in lipid bilayer membranes

B42   R. Benz and U. Zimmermann   64
       Reversible electrical breakdown of lipid bilayer mem-
       branes

B43   U. Zimmermann and R. Benz   65
       Dependence of the electrical breakdown voltage on the
       charging time in Valonia utricularis

B44   K. Schmidt and R. Benz   66
       Interaction of burn toxin with liver cells and artifi-
       cial lipid bilayer membranes

B45   U. Zimmermann, G. Pilwat and H. Schnabl   67
       Development of a new Coulter counter system: Measure-
       ment of the volume, internal conductivity, and dielec-
       tric breakdown voltage of a single guard cell protoplast
       of Vicia faba and tumor cells

B46   J. Vienken, H. Schnabl, U. Zimmermann and H. Ziegler   68
      Guard and mesophyll cell protoplasts of Vicia faba:
      Plasmalemma ultrastructure

B47   N. Hodapp, W. Welte, D. Walter and W. Kreutz   69
      X-ray diffraction and electronmicroscopic studies of
      thylacoid membranes

B48   R. Tiemann, D. DiFiore and H.T. Witt   70
      On the transmembrane electrical potential difference
      in chloroplasts studied by electrochromism

B49   R. Tiemann, G. Renger and P. Gräber   71
      Study of the electron- and proton transport in inside-
      out thylakoids

B50   E. Schlodder and H.T. Witt   72
      Initial kinetics of ATP-synthesis and of conformational
      changes in the chloroplast ATPase studied by external
      electric field pulses

B51   G.H. Schatz, E. Schlodder, M. Rögner and P. Gräber   73
      Rapid kinetics of ATP-synthetis in chloroplasts by acid
      /base transition

B52   H.-L. Huber and B. Rumberg   74
      Demonstration of electrical surface charge effects at the
      inner side of the thylakoid membrane

B53   H. Felle   75
      Ammonium transport in the plasma membrane of Riccia flui-
      tans

B54   C. Hertel and D. Marmé   76
      Effect of anti-microtubule herbicides on Ca^{2+} transport
      and energy transduction in plant mitochondria

B55   J. Funk, I. Wutschel, W. Welte and W. Kreutz   77
      Evaluation of the electrondensity profile of the frog
      ROS-disk membrane in vivo using X-ray diffraction

B56   P.P.M. Schnetkamp, U.B. Kaupp and W. Junge   78
      Surface potentials in intact cattle rod outer segments

B57   R. Uhl, N. Semple, J. Pasternak, T. Borys and
      E.W. Abrahamson   79
      On the protein composition of bovine rod outer segment
      disk membranes

B58   R. Uhl, T. Borys and E.W. Abrahamson   80
      Structural consequences of magnesium ATP-ase activity
      in bovine rod outer segment disk membranes

B59   J.-D. Spalink   81
      Flash spectroscopic studies on electrophysiologically
      intact bovine retinae

B60   J. Bernhardt and E. Neumann   82
      Chemical control of membrane transport by the acetyl-
      choline receptor system

B61   E. Neher and F. Conti   83
      Discrete current fluctuations produced by single K^{+}-
      channels in the squid axon membrane

B62   W. Stühmer and F. Conti   84
      The effect of high extracellular potassium on the kine-
      tics of potassium conductance of the squid axon membrane

B63   W. Finger and J. Dudel   85
Effect of GABA on non-synaptic potassium channels in
crayfish muscle

B64   H. Stettmeier, W. Finger and J. Dudel   86
Excitatory synaptic current noise in crayfish muscle
fibres

B65   W. Carius   87
The Schwann cell sheath modifies the dielectric properties of squid axon preparations

B66   J.E. de Peyer and J.W. Deitmer   88
Ion-selectivity of voltage- and of mechano-sensitive
membrane channels in Stylonychia

B67   J. Gödde and U. Thurm   89
Functional separation of subcellular sensitive elements
by selective adaptation in a ciliary mechanoreceptor cell

B68   G. Burckhardt and H. Murer   90
Optical monitoring of membrane potential changes as a
tool to study sodium-dependent substrate transport in
brush border vesicles from kidney proximal tubules

B69   P.G. Wood and U. Rempel-Rossleben   91
Is the spontaneous leak found in the resealed ghost related to the Gardos effect?

B70   J.P. Seher and G. Adam   92
Transformed cells differ from normal cells by the topo-
logical arrangement of the outer cell membrane

B71   U. Steiner and G. Adam   93
Fluorescence polarization of 3T3 and SV40-3T3 cells thinly spread on their growth substrate

B72   D. Tobüren   94
The effect of bleomycin on V 79-spheroids

B73   H.-P. Beck   95
Disturbance of proliferation kinetics of L-cells by
^{3}H-thymidine labelling

B74   W. Strobelt, Th. Vömel and D. Platt   96
Scanning electron microscopic measurement of human red
blood cell diameter alterations as a function of cell age

B75   J. Schreiber, W. Greulich and W. Lohmann   97
Influence of reducing and oxidizing substances on healthy
and "leukemic" blood and its fractions

B76   W. Greulich, J. Schreiber and W. Lohmann   98
Interaction between ascorbic acid and Cu-proteins: Atomic
absorption and ESR measurements on erythrocyte ghosts and
blood plasma

B77   G. Pilwat, J. Vienken and U. Zimmermann   99
Organ specific application of drugs by means of cellular
capsule systems

B78   K.-H.C. Standke   100
Investigation of biophysical parameters in deep freezing
experiments with plant cells in order to achieve a high
rate of surviving frost sensitive cells

B79   M.H. Weisenseel   101
Natural electric currents traverse growing plant cells
and tissues

B 80   L. Edelmann   102
        Selective accumulation of alkali-metal ions at cellular
        protein sites without ion pumps

Section C: Neurobiology and Cybernetics   103

C1   H. Stieve, W. Schröder, M. Bruns and I. Claßen-Linke   104
        Exchange of solutes between superfusate and the extra-
        cellular compartment adjacent to the photosensory mem-
        brane of Limulus and crayfish photoreceptor

C2   W. Schröder, D. Frings and H.J. Heinen   105
        Laser-microprobe-mass-spectrometry reveals two classes
        of shielding pigment in the Astacus retina: One binding
        large amounts of Ca, the other binding Na and K.

C3   P.J. Bauer and H.G. Smith, Jr.   106
        Light-induced calcium release from bovine disk vesicles

C4   U.B. Kaupp, P.P.M. Schnetkamp and W. Junge   107
        The kinetic behaviour of rapid calcium release in cattle
        rod outer segments: The metarhodopsin I/metarhodopsin
        II-transition is involved in the trigger mechanism

C5   B. Walz   108
        Calcium-containing and calcium-accumulating structures
        in photoreceptor cells of the leech Hirudo medicinalis

C6   J. Wulf   109
        Effects of extracellular Ca^{++} on the response of photo-
        receptor cells of Hirudo medicinalis

C7   D. Emeis and K.P. Hofmann   110
        Time course and temperature dependence of the light in-
        duced alkalisation effect in bovine rod outer segments

C8   K.P. Hofmann and D. Emeis   111
        Evidence for light induced lateral contraction of the
        disk membrane by small bleachings of rhodopsin (a more
        detailed interpretation of the "P-signal")

C9   W. Ehrhardt   112
        Electrical properties of the external limiting membrane
        in the receptor layer of the frog retina

C10   W.S. Jagger   113
        Mechanism of photoresponse generation in isolated frog
        rod outer segments

C11   H. Hatt and U. Bauer   114
        Chemoreception in the crayfish Orconectes limosus

Section D: Radiation Biophysics   115

D1   B. Mukherjee   116
        A simple irradiation facility with protons

D2   O. Merwitz   117
        The autoradiolytic and the gamma-induced demethylation
        of thymine in aqueous solution

D3   M. Frankenberg-Schwager, D. Frankenberg, D. Blöcher
        and C. Adamczyk   118
        Measurement of irreparable double-strand breaks in the
        DNA of eukaryotic cells

D4   G. Tisljar-Lentulis, P. Henneberg and L.E. Feinendegen   119
     The oxygen enhancement ratio for single- and double-
     strand breaks induced by tritium incorporated in DNA
     of cultured human kidney cells

D5   E. Dikomey and H. Jung   120
     Repair of DNA strand breaks in CHO cells after low doses
     of X-rays

D6   R. Zblewski and H. Rink   121
     UV excision repair in lens epithelial cells during aging
     in vitro

D7   V. Kasche, U. Frixen, J. Maas, K. Probst and P. Zipfel   122
     Radiation induced protein-DNA crosslinks in eucaryotic
     cells

D8   R. Seidler and W. Köhnlein   123
     The effect of long-wavelength UV on BrUra substituted
     E. coli: Correlation between survival, strand breaks,
     and BrUra-incorporation

D9   H. Dertinger and D.F. Hülser   124
     Correlation between enhanced radioresistance of spheroid
     cells and cell coupling

D10 B. Kessler-Rosbach, H.-J. Meyer-Teschendorf and H. Rink   125
     Influence of X-irradiation on the rosette-forming capa-
     city of human lymphocytes

D11 C. Baumstark-Khan and H. Rink   126
     Observations in synchronized yeast cells (Sacch.uv.)
     during X-ray-induced giant cell formation

D12 J. Kiefer and I. Wienhard   127
     Repair and recovery in irradiated yeast - the effect of
     protein synthesis inhibitors on liquid holding recovery
     in X-irradiated yeast

D13 J. Luggen-Hölscher, S. Rase and J. Kiefer   128
     Repair and recovery in irradiated yeast. Mutation induc-
     tion by ionizing radiation: Influence of oxygen and LET

D14 K.J. Weber, H. Waller and J. Kiefer   129
     Repair and recovery in irradiated yeast. Excision repair
     of yeast DNA after exposure of UV (254 nm)

D15 F. Zölzer and J. Kiefer   130
     Repair and recovery in irradiated yeast. Lethal and
     mutagenic effects of 254 nm- and 313 nm-radiation on
     yeast strains of different repair capabilities

D16 F. Schöpfer, S. Rase, E. Schneider and J. Kiefer   131
     Repair and recovery in irradiated yeast. Survival and
     mutation induction by heavy ions in yeast cells of dif-
     ferent radiosensitivity

D17 W. Grundler   132
     The kinetic of yeast cells within the first four genera-
     tions after irradiation with ionizing particles

D18 L. Schachinger, H. Klöter, M. Michailov and Ch. Schippel   133
     The possible role of c-AMP in the irradiation induced
     contraction of nerve muscle preparation

D19 A. Kindt, E.L. Sattler and A. Schraub   134
     Irradiation and gastric emptying

D20   C. Streffer, M. Molls, N. Zamboglou and D. van Beuningen   135
The effect of neutron- and X-irradiation on the development of preimplanted mouse embryos

D21   E. Werner, W. Pohlit and K.P. Schalk   136
Immediate effects of ionizing radiation in rabbits

D22   U. Schrader-Reichhardt, B. Markus and B. Papenberg   137
Interaction of 40,5°C-hyperthermia with 200 kV X-rays
at two different dose-rates

D23   I. Schmitz-Feuerhake and E. Muschol   138
Late effects from low LET radiation in the low dose
range - the experience of Hiroshima and Nagasaki

D24   M. Matthies, H.G. Paretzke and W. Jacobi   139
A non-linear model of the global carbon cycle for the
assessment of long-term effects of CO_2 and C-14 due to
various energy scenarios

Section A

MOLECULAR BIOPHYSICS

7-CIS RETINAL, AN ARTIFICIAL CHROMOPHORE OF BACTERIORHODOPSIN

K.-D. Kohl and W.Sperling
Institut für Neurobiologie der Kernforschungsanlage Jülich GmbH
Postfach 1913
D - 5170 Jülich

Bacteriorhodopsin (BR) is a chromoprotein and the pigment of the purple
membrane of *Halobacterium halobium*. The naturally occurring chromophore
of BR consists of a mixture of trans and 13-cis retinal. The retinal-
free protein is called Bacterioopsin (BO) and is regenerable with both
13-cis or trans retinal.
Because of the structural similarity to trans retinal, we expected also
7-cis retinal to combine with BO.

trans   7-cis
                    (the ring is perpendicular to the
                     plane of the conjugated chain)

Indeed, BO reacts with 7-cis retinal to yield a pigment with a λ_{max} of
460 nm. In this equilibrium of 7-cis retinal with BR_{7-cis}, the free
7-cis retinal is favored. This is in contrast to the 13-cis/trans BR-
system, where the corresponding equilibrium is nearly completely on
the side of the BR. A shoulder appearing at 560 nm we assigned to 7,13-
dicis BR, which equilibrates with 7-cis BR.
The 7-cis/7,13-dicis BR-system is photochemically and thermally
irreversibly convertible to the native BR-system.

ON THE DETERMINATION OF THE ABSORPTION SPECTRA OF SOME 11-CIS-RETINAL CONFORMERES BY POLARISATION SPECTROSCOPY USING EXTREMELY THIN CRYSTAL PLATES

G.Drikos, H.Rüppel

Max Volmer Institut für Physikalische Chemie und Molekularbiologie, Technische Universität Berlin

W.Sperling

Institut für Neurobiologie, Kernforschungsanlage Jülich (KFA)

The 11-cis-retinal is the chromophoric group of all known visual pigments. It's unique importance in animal kingdom corresponds to that of the chlorophyll in vegetable kingdom.

11-cis-retinal has special properties such as the anomalous temperature dependence of absorption in solution, but also in vitro. This feature can be interpreted by means of a conformere equilibrium.

Some time ago a 12-s-cis conformation was actually discovered in crystallized 11-cis-retinal by Karle using X-ray analysis. Meanwhile we succeeded in crystallization of 11-cis-retinal in plates which were thin enough ($<1\mu$m) to measure polarisation spectra in the microspectrophotometer. Thereby it turned out that two different types, we called A and B, of 11-cis-retinal crystals exist, which are different conoscopical as well as significantly in the polarisation spectrum.

Another check-up of the X-ray structure confirmed the structure found by Karle for type A. In comparison with it in the B-type another crystal cell and most likely the 12-s-trans conformation is present.

The polarisation spectra of the 11-cis conformation are inter-
preted by means of both crystal structures (an X-ray analysis of type B is under investigation) and beyond this are compared with the spectra in solution.

Resonance CARS Spectroscopy of Bacteriorhodopsin

in the Nanosecond Range.

J. Tretzel and F. W. Schneider
Institut für Physikalische Chemie der Universität Würzburg,
Marcusstr. 9/11, 8700 Würzburg, GFR

CARS (Coherent Anti Stokes Raman Spectroscopy) is a relatively new nonlinear optical spectroscopic method which shows orders of magnitude higher Raman efficiency than conventional Raman spectroscopy. It is particularly suitable for the study of short lived intermediates in the nanosecond range and also for recording Raman spectra of strongly fluorescent compounds. High time resolution is achieved by the use of pulsed lasers of high power.

We report the first resonance CARS spectra of light and dark adapted bacteriorhodopsin in the nanosecond time range. Differences between the spectra obtained by the conventional resonance Raman and the present resonance CARS method are discussed. A new band has been found for the dark adapted bacteriorhodopsin which may be assigned either to the 13 cis form or to a new intermediate whose risetime is in the subnanosecond range.

KINETIC INFRARED SPECTROSCOPY ON THE PHOTOCYCLE

OF BACTERIORHODOPSIN

W. Mäntele, F. Siebert, W. Kreutz

Institut für Biophysik und Strahlenbiologie
der Universität Freiburg im Breisgau
Albertstrasse 23, D-7800 Freiburg, W. Germany

We have applied the method of flash photolysis with infrared measuring light to the photochemical cycle of bacteriorhodopsin (BR). This method developed by us yielded first results in the study of the retinal-protein-interaction of rhodopsin.

Because of the stability of bacteriorhodopsin and its cyclic reaction a large spectral range could be covered with one sample and kinetic signals of one sample could be averaged. In the spectral region from 1100 cm^{-1} to 1800 cm^{-1} kinetic difference spectra of samples could be obtained at a time resolution of 100 microseconds.

These difference spectra show a strong absorbance decrease at 1525 cm^{-1} which can be assigned to the C=C stretching vibration of BR_{570} in good agreement with Raman spectroscopic measurements. This also holds for the fingerprint region.

A strong absorbance increase at 1555 cm^{-1} can possibly be assigned to the C=C stretching vibration of the N_{530}-species, which in the infra-red region should absorb at shorter wavelengths than BR_{570}. Its rise-time is different from that of the 1525 cm^{-1} absorbance decrease.

In the Schiff base region the difference spectrum shows a small but broad absorbance decrease from 1630 cm^{-1} to 1680 cm^{-1} which cannot be interpreted in terms of a BR_{570} species protonated in the usual sense. We suggest - as already done for rhodopsin - a protonation via a hydrogen bond.

KINETIC INFRARED INVESTIGATION ON THE
PHOTODISSOCIATION OF SPERM WHALE
CO-MYOGLOBIN

F. Siebert, W. Mäntele, W. Kreutz

Institut für Biophysik und Strahlenbiologie
der Universität Freiburg im Breisgau
Albertstrasse 23, D-7800 Freiburg, W. Germany

With our method of kinetic infrared spectroscopy we investigated the photodissociation of sperm whale CO-myoglobin in the region of the CO-vibration band. With the infrared beam at a wavelength of the CO-vibration of bound CO we observed flash induced signals whose time courses correspond to the photodissociation of CO-myoglobin and its rebinding as measured at the soret band. The amplitudes of the flash induced signals reproduced the static measured difference spectrum of CO-myoglobin vs. myoglobin, with a maximum at 1944 cm^{-1} and a shoulder at 1932 cm^{-1}. This shoulder has generally been taken as an evidence for a second binding site for CO in myoglobin. It could also be that there exists an equilibrium between two types of binding at the same binding site. We hoped to get information about this problem by measuring the spectral dependence of the rebinding kinetics. Within the experimental errors no dependence could be detected. The most plausible explanation for this would be a fast equilibration between two types of binding at the same binding site. Photolysed CO dissolved in water cannot be detected due to the large intermolecular interaction with the water molecules. At the time scale of our measurements we have no evidence for an intermediate binding site of CO, since no band developed (no absorbance increase was measured), which could be attributed to the CO-vibration.

SOME THEORETICAL AND PRACTICAL ASPECTS OF ENZYME IMMOBILISATION AT THE EXAMPLE UREASE/ACTIVATED ALUMINIUMHYDROXIDE

P. Grunwald, W. Gunsser, E. Schober
Institut für Physikalische Chemie
der Universität Hamburg
D-2000 Hamburg 13, FRG

The immobilisation of enzymes is a well-known technique (enzyme engineering) by which proteins are made water-insoluble. In several cases it was possible to obtain new fundamental knowledge in enzyme chemistry from studies on the reaction behaviour of immobilised enzymes (1). We have carried out investigations on urease covalently attached to carriers of the type $Al_n(OH)_{3n-1}-OOC-C_6H_4-NH_2$ (2) by an azo linkage (3). Normally, the activity of enzymes is reduced by immobilisation procedures, but this disadvantage can be overlooked because of the possibility of reusing the enzyme. The residual activity, too, is a function of the number of attachment points between protein and carrier per unit area. For the carrier material applied by us, low 'n' values mean high p-aminobenzoicacid concentrations and the activity of enzyme preparations increases with increasing value of 'n'. Though enzymes, that are bound more rigid to a solid support have smaller activities, they are, as our experiments show, of better durability, if used for continous urea hydrolysis during several weeks, even if the reaction is performed at higher temperatures.

Of theoretical interest is the temperature dependence of the urea hydrolysis in presence of immobilised urease. If $'n' \leqslant 15$, we observe two successive reaction ranges at any given temperature that come out even more distinct the stronger the enzyme is bound to the carrier surface. We attribute this effect to the existence of two active sites E' and E" in the enzyme urease with different substrate affinity. The activation energies are about 20 kJ/mole for E' and 40 to 50 kJ/mole for E". Moreover, E' and E" show different temperature stabilities. The optimum temperatures are $60^{\circ}C$ (E') and about $47^{\circ}C$ (E"). Our results confirm earlier assumptions about the existence of more than one active sites for urease, that were based on other experiments (4).

1 Mosbach, K. (1976) FEBS Lett., 62 Suppl. E 80 - E 95
2 Grunwald, P., Gunsser, W. (1978) Naturwiss. 65, 60
3 Weetall, H.H., Hersh, L.S. (1969) Biochim.Biophys.Acta 185,464-465
4 Lynn, K.R., Yankwich, P.E. (1964) Biochim.Biophys.Acta 81, 533-547

KINETIC OF UREASE-CATALYSED UREA HYDROLYSIS IN THE PRESENCE OF ALKALINE EARTH HALIDES

P. Grunwald, W. Gunsser
Institut für Physikalische Chemie der Universität Hamburg,
D-2000 Hamburg 13, FRG

In order to obtain informations about the reaction mechanism of enzyme-catalysed reaction, a simple kinetic method exists, that shall be demonstrated here with urease as an example, the reaction behaviour of which is not yet known. The procedure is based on the conversion of the substrate molecule S into a charged state (eq.1),whose reaction with active site of the enzyme is then investigated (eq.2). In the case in question we made use of the complex formation between urea and alkaline earth halides (M_eX_2).

$$S + Me^{++} \longrightarrow (SM_e)^{++} \qquad (1)$$

$$(SM_e)^{++} + E \rightleftharpoons (SM_e)^{++}\text{-}E \longrightarrow E + P + M_e^{++} \qquad (2)$$

The reaction velocity was measured at different salt concentrations; the added salts were $MgCl_2$, $CaCl_2$, $SrCl_2$ and $BaCl_2$. M_eX_2 varied between 10^{-4} and 10^{-6} mole/l, a concentration range, where the Debye-Hückel theory is valid and where a possible influence of X^- on the enzyme reaction is negligible (1). The interpretation of the reaction rates as a function of ionic strength I was made by applying the Debye-Hückel limiting law and the theory of transition state to reactions in solutions, which leads to the following equation

$$\log k = \log k_0 + 1.02 \cdot z_S \cdot z_E \sqrt{I} \qquad (3)$$

k_0 is the reaction rate at infinite dilution, z_S and z_E are the charges of substrate and active site of the enzyme. Indeed, all experimental points fall close to a straight line given by eq.3. Moreover, urease activity is only affected by the ionic strength and not by specific properties of added salts. From the negativ gradient of the experimental line it must be concluded, that the active site of urease carries a negative charge.

The influence of alkaline earth halides on urease is confirmed by measuring the temperature dependence of this reaction; this being performed in small temperature intervals ($\Delta T \sim 1^\circ C$) between $3^\circ C$ and $80^\circ C$. The anomalous temperature behaviour of urease (2) is distinctly intensified with increasing ionic strength. Further the presence of MeX_2 results in a considerable stabilisation of urease at higher temperatures. The optimum temperature increases from $60^\circ C$ to $72^\circ C$ if the reaction solution contains 10^{-4} mole/l $SrCl_2$.

Kistiakowski, G.B., Mangelsdorf, P.C., Rosenberg, A.J.,Shaw, W.H.R. (1952) J. Am. Chem. Soc. 74, 5015-5020
Talsky, G. (1971) Angew. Chem. 83, 553-560

ION ATMOSPHERE RELAXATION OF SIMPLE ELECTROLYTES AND POLYIONIC BIOPOLYMERS IN ELECTRIC FIELDS.

Dieter Schallreuter and Eberhard Neumann
Max-Planck-Institut für Biochemie, D-8033 Martinsried, BRD

A recently developed Electric-Field-Jump apparatus (1) with high sensitivity and time resolution permits the measurement of very rapid dynamic phenomena in electrolytic and dipolar systems. Conductivity relaxations induced by rectangular electric field pulses (1 to 100 kVcm^{-1}), of simple alkali metal salts, linear oligo- and polyelectrolytes and polyionic proteins can be analyzed in terms of the Debye ion cloud relaxation and of the Wien field-dissociation effect of ion clouds and ion pairs. The relaxation data suggest that even in simple electrolytes like NaCl there is a kind of larger interaction domains. These dynamic, perhaps lattice-like, interaction clusters have larger inertia than the individual ions; the ion density in such a cluster depends on the concentration. Furthermore, so-called strong electrolytes like NaCl display a small field-dissociation effect, suggesting that at higher concentrations (0.1 M) even NaCl in aqueous solution is not completely dissociated (2). Finally, the relaxation spectra of polyionic systems are particularly informative. For instance, polynucleotides exhibit a roughly two-phasic relaxation suggesting a substructure of the ionic atmosphere: a Bjerrum type multiple association of counterions near the polyion; this region of ion pairing is surrounded by the classical Debye-Hückel ionic atmosphere. The consequences of these data for models of polyelectrolyte catalysis and surface diffusion in biopolymers and biomembranes are discussed.

(1) Schallreuter, D., Rohner, J., and Neumann, E.,
    Rev. Sci. Instr. (1979), submitted.

(2) Fuoss, R.M., Proc. Natl. Acad. Sci. USA 71, 4491-4495
    (1974); 75, 16-20 (1978); Barthel, J., Ber. Bunsenges.
    Phys. Chem. 83, 252-257 (1979).

HIGH-RESOLUTION TEMPERATURE-JUMP MEASUREMENTS IN BIOCHEMISTRY

Carl-Roland Rabl

Max-Planck-Institut für Biochemie

D-8033 Martinsried/München, BRD

Spectrophotometric relaxation methods offer an available time range from the ns- to the sec-region. They require much less material than Stopped Flow techniques but need also a very high signal resolution which is especially important in biochemical applications. Signal resolution is limited by photon noise in the short time range and by light flux instabilities in the long time range, but also by the transient response of the apparatus and transient distortions due to the perturbation method, by nonlinearities of the photodetectors, improper signal balancing, etc. A careful analysis of these errors and application of appropriate technical measures resulted in the construction of a temperature-jump apparatus with a differential signal resolution of up to 10 ppm. The effective time resolution has been improved by convolution analysis of the measured signal. In the long time range, cooling of the sample following the temperature-jump can introduce serious errors and has been corrected for by an analog deconvolution circuit. [1, 2, 3]

Examples of applications are given, such as the study of competive binding of acetylcholine and calcium to the acetylcholine receptor, using Murexide as an Ca-indicator [4, 5].

Accessories, such as for combined Stopped-Flow Temperature-Jump, can be used for extending the applications of the apparatus [6].

References:

1) C.R.Rabl, in: Technische Biochemie, ed. H.J.Rehm, Dechema Monographies Vol.71, Verlag Chemie, Weinheim 1973, pp. 187-203

2) R.Rigler, C.R.Rabl, and T.M.Jovin, Rev.Sci.Instr. 45 (1974),580 - 588

3) C.R.Rabl, in: Techniques and Applications of Fast Reactions in Solution, eds. W.J.Gettins and E.Wyn-Jones; Reidel, Dortrecht 1979

4) E.Neumann and H.W.Chang, Proc.Nat.Acad.Sci.USA 73 (1976), 3994-3998

5) C.R.Rabl and E.Neumann, in preparation

6) together with R.Rigler, T.M.Jovin, et al.

DEMONSTRATION OF G·U WOBBLE BASE PAIRS BY RAMAN AND IR SPECTROSCOPY

H.Klump, Th.Ackermann, V.Gramlich, [*]Th.Knäble,
E.D.Schmid, [*]H.Seliger and J.Stulz.

Inst. Physik. Chemie, Universität Freiburg, W.Germany,
and [*]Sektion Polymere, Universität Ulm, W.Germany.

When guanine and uracil form hydrogen bonds in the
pairing scheme first proposed by Crick [1] one would
expect that poly (A,G) will form an unperturbed double
helix with poly U at room temperature in a dilute
electrolyte solution (0.1 M NaCl). We have demonstrated
by Raman and IR spectroscopy that the secondary structure
of poly (A,G) • poly U is very similar to the structure
of poly A • poly U; only the thermal stability of the
double helix seems slightly lower than the stability of
poly A • poly U, Whereas the average helix length is
unaffected by the dispersed G • U base pairs. From our
input ratio of guanine and adenine we estimate that
about every fourth base pair is a wobble pair.

1.   Crick, F.H.C. (1966) J.Mol.Biol. 19, 548-555

A NEW MODEL FOR THE ACTIVATION AND INACTIVATION OF NEOCARZINOSTATIN, AN
ANTITUMOR PROTEIN

W. Köhnlein, R.S. Lewis and G. Jung.
Institut für Strahlenbiologie der Westfälischen Wilhelms-Uni-
versität, Münster, Germany.

Neocarzinostatin (NCS), an antibiotic protein exhibits therapeutic
activity against a variety of tumors, most notably acute leūkemia in
man. The protein has a MW of 10,700 and the known amino acid sequence
includes four cysteines connected by two disulfide bonds. The in
vitro substrate for NCS is double stranded DNA, and the drug makes
3',5' phosphate bounded single strand breaks.
Sulfhydryl compounds specifically activate and inactivate NCS as measured by strand scission of T2 DNA. This effect, together with the
failure of attempts to alkylate cystines in NCS and evidence for
strained disulfides in the molecule, leads us to propose a new model
of activation and inactivation:

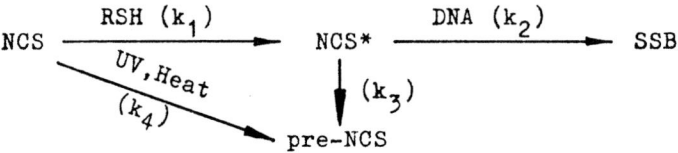

We suggest that NCS cystines are reduced by sulfhydryl reagents to
produce a short-lived active form which may react with DNA or proceed
irreversibly via a conformational change to an inactive form(pre-NCS).
Reduction potential dependent activation by 2-mercaptoethanol and di-
thiothreitol and the inhibition of strand breakage by the possible
blockage of free NCS SH groups with an excess of disulfide supports
the assumption that free SH groups are necessary for activity. Three
parameters of NCS activity - inactivation rate, single strand break
plateau position, and initial SSB rate - have been measured for various sulfhydryl concentrations, and the observed results agree well
with values expected from a simplified mathematical treatment of the
model, thus supporting the assumptions made.
A similar analysis of date obtained for UV irradiation and heat treatments indicate the presence of an alternate pathway from NCS to pre-
NCS which does not pass through an active intermediate.

GENERATION OF PRE-NEOCARZINOSTATIN AND ITS ANTAGONISTIC EFFECT ON
NEOCARZINOSTATIN-INDUCED DNA STRAND SCISSION

G. Jung, R.S. Lewis and W. Köhnlein
Institut für Strahlenbiologie der Westfälischen Wilhelms-
Universität, Münster, Germany.

Pretreatment of the antitumor protein neocarzinostatin (NCS) with
heat, UV or room light, and sulfhydryl reagents inactivates the drug,
as measured by the cessation of phage T2 DNA strand scission in vitro.
The inactive forms obtained are identical with pre-neocarzinostatin
(pre-NCS) on the basis of isoelectric focusing, molecular weight
determination, and changes in circular dichroism (CD) spectra.
Addition of pre-NCS to incubations of T2 DNA and NCS inhibits strand
breaking activity in a ratiodependent NCS/pre-NCS manner, indicating
a competition between the two proteins for a limited number of DNA
binding sites. A quantitative analysis demonstrates that pre-NCS
binds at least 3 to 4 times more strongly than NCS at 37° C, and the
two bind roughly equally at 54° C. These different ratios could be
explained by proportionately greater temperature effects on pre-NCS
fine structure which may occur as a result of its looser overall
conformation.
The clear inhibition of NCS strand breaking activity in vitro by pre-
NCS is in good agreement with effects seen in vivo. Although competition between the two isomers for a limited number of cell surface
receptors cannot be ruled out, the correlation between antagonistic
effects in vivo and in vitro clearly suggests that binding of NCS to
DNA is an essential aspect of the drug's antitumor action in vivo.
An attempt to selectively reduce the magnitude of the antagonistic
effect in tumor cells, for example by lowering the binding affinity
of pre-NCS for DNA through local heating of solid tumors, could be
of value in raising the therapeutic effectiveness of NCS.

THE PRODUCTION OF STRAND BREAKS BY THE INTERACTION OF BLEOMYCIN WITH
SUPERCOILED COL E1 DNA: AN UNEXPECTED CONCENTRATION DEPENDENCE

F. Zimmermann and W. Köhnlein.

Institut für Strahlenbiologie der Westfälischen Wilhelms-
Universität, Münster, Germany.

The glycopeptide bleomycin (BLM) isolated from Streptomyces vertizillus
has been shown to possess antitumor activity in many systems. We have
been able to show that there exists a similarity between the effect of
BLM on DNA and high LET radiation. In both cases a direct production
of double-strand breaks is observed.

For a better understanding of the radiomimetic action of BLM we have
investigated in detail the strand-breaking mechanism of the drug on
supercoiled Col E1 DNA.

The use of supercoiled DNA and agarose gel electrophoresis as detection system for reaction products (circular and linear DNA molecules)
allows identification and quantification of single and double strand
breaks simultaneously in the same experiment.

We found a linear time dependence at all BLM concentrations used for
the number of single and double strand breaks as well. The reaction
rates, however, do not show a simple correlation with the bleomycin
concentration. An oscillating amount of strand breaks with increasing
values of r (r = $[BLM]/[DNA]$) was found. Maxima of strand break ability
were observed at $\triangle r$-values of $2,5 \cdot 10^{-3}$ BLM/DNA-base which is equivalent to an additional intercalation of about 28 BLM-molecules/DNA-
molecule. At those maxima no alkali labile bonds were produced as can
be judged from the ratio breaks at pH 12/breaks at pH 7,5. Through the
intercalating process the free energy of the DNA molecule is raised
until one supertwist is removed and the DNA assumes a conformation of
lower free energy. Upon further intercalating BLM-molecules this
process is repeated. Strand-breaking ability seems to be high at conditions of high free energy. These results led us to assume that the
number of strand breaks produced by the intercalating BLM strongly
depend on the supercircular structure of the DNA.

ELECTRON SPIN RESONANCE STUDY OF COBALT LEGHAEMOGLOBIN

Christahl, M., Gersonde, K., Raap, A. and *Appleby, C.

Rheinisch-Westfälische Technische Hochschule Aachen
D-5100 Aachen, Melatener Str. 213, Germany

* Commenwealth Scientific and Industrial Research Organization
Canberra, A.C.T. 2601, Australia

Leghaemoglobin is an oxygen-binding haemoprotein which occurs in the bacteroid-containing cells of soybean nodules [1,2]. This haemoglobin is characterized by a very high affinity for O_2 ($p_{O_2(1/2)}$ = 0.04 mmHg, pH = 7.4, 25° [2]. From the nodules two monomeric components a and c can be prepared [1]. In our experiments the mixture of leghaemoglobins a and c is used.

Both, the deoxy and the oxy state of the leghaemoglobin does not show electron spin resonance (ESR). However, substitution of the haem iron by cobalt leads to ESR-active oxy and deoxy haemoglobins with S = 1/2. The ESR-spectrum of oxy cobalt leghaemoglobin reflects rhombic symmetry with g_{xx} = 2.077, g_{yy} = 1.987 and g_{zz} = 2.004. The assignment of the g-values to the main directions is possible by the comparison with the respective nitrosyl iron haemoglobin spectra. The use of the second derivative spectra allows the determination of the cobalt hyperfine constants in the three directions (a_{xx}(Co) = 1.56 mT, a_{yy}(Co) = 0.90 mT, a_{zz}(Co) = 0.96 mT).

The ESR spectrum of deoxy cobalt leghaemoglobin is of nearly axial symmetry with $g_{xx} \approx g_{yy}$ = 2.312 and g_{zz} = 2.032. Only in g_{zz} a hyperfine structure is resolved, indicating the interaction of the unpaired electron with both nuclei, i.e. cobalt and imidazole nitrogen. The cobalt hyperfine splitting is a_{zz}(Co) = 7.92 mT, that of the triplets is a_{zz}(N_{ε}) = 1.77 mT. The deoxy state of this haemoglobin seems to be pH-insensitive, especially with respect to the hyperfine structure in g_{zz}. On the other hand the oxy state shows a change of the hyperfine constants in g_{zz} varying from low to high pH. This pH-dependence of the hyperfine constants reflects a variation of the spin density at the cobalt nucleus with pH.

From this result we expect that cobalt leghaemoglobin shows a pH-dependent oxygen-binding, i.e. a Bohr effect. To support this, O_2-binding experiments are under study.

1 Appleby, C.A., J. Biol. Chem. 251 (1976) 6090.

2 Wittenberg, J.B., Bergersen, F.J., Appleby, C.A., Turner, G.L.,
   J. Biol. Chem. 249 (1974) 4057.

INTERACTIONS OF ATP PHOSPHATE GROUPS WITH AMINES AND DIVALENT METAL IONS: ^{31}P-NMR STUDIES

H.Sapper, W.Gohl, W.Lohmann
Institut für Biophysik, Strahlenzentrum der Universität Gießen,
Leihgesterner Weg 217, D-6300 Gießen

The mechanism of uptake, storage, and release of several biogenic amines acting as neuronal transmitters (as well as the molecular effect of some psychoactive catecholamines) seems to be determined essentially by the interaction of the amines with adenosine-5'-triphosphate (ATP). ^{31}P-NMR spectroscopy of the ATP phosphorous groups has been used for determining the uptake of amines by chromaffin granules[1] and investigating qualitatively the involvement of the different phosphate groups in the association with catecholamines[2] or divalent metal ions[3].

The strength, loci, and, to a certain extent, the stoichiometry of binding between amines, divalent metal ions, and ATP have been evaluated quantitatively by the use of the pH-dependence of the phosphorous resonances of ATP. The ^{31}P-NMR observations complete, thus, the ^{1}H-NMR and IR studies performed[4] which indicated an electrostatic amine phosphate attraction modified by certain substituent effects. The association of biogenic (as well as artificial) amines with ATP seems to prefer the terminal phosphate group resulting in association constants K between 10 and 30 M^{-1} (25°C). Divalent metal ions (e.g. Mg^{2+}, Ca^{2+}) seem to bind also to the ß-phosphate group of ATP ($K \approx 10^{3}$ M^{-1}) interfering, thus, with the amine association at the terminal phosphate group. This competition of the metal ions with amines at the phosphate binding sites of ATP might be responsible for the release of amines from their storage granules.

1) Casey,R.P.,Njus,D.,Radda,G.K.,Sehr,P.A.:Biochem.16(1977)972-977
2) Granot,J.: FEBS Lett. 88(1978)283-286
3) Cohn,M.,Hughes,T.R.: J.Biol.Chem. 237(1962)176-181
4) Sapper,H.,Gohl,W.,Matthies,M.,Haas-Ackermann,I.,Lohmann,W.:
   Cell.Mol.Biol.(submitted for publication)

INVESTIGATIONS ON THE INTERACTION OF NADH WITH Cu(II): ESR AND OPTICAL ABSORPTION SPECTROSCOPY

W.Wallbott, H.Neubacher, W.Lohmann
Institut für Biophysik, Strahlenzentrum
Justus-Liebig-Universität Giessen
6300 Giessen, FRG

The interaction of the transition metal ion Cu(II) with NADH and its constituents has been investigated by means of electron spin resonance (ESR) and optical absorption spectroscopy. It has been shown by the decrease of the 340 nm absorption band of NADH that NADH is oxidized rapidly in the presence of Cu(II) ions. One Cu(II) ion seems to be responsible for more than one oxidation. In aquoeous solution, Cu(I) is reoxidized very rapidly to Cu(II). Thereafter, Cu(II) seems to form slowly a complex with NAD, as can be seen by the shift of the d-d band of the aquo complex from 800 nm to 690 nm. The molar extinktion coefficient increases concomitantly. This complex, then, seems to be stable for several weeks. The coordination sites of the NAD molecule seem to be N(7) of the adenine residue and O of the adjacent phosphate group. This can be concluded from the ESR and optical results obtained with Cu(II)-AMP complexes. Since both of these complexes exhibit very similar g-values and d-d-transition energies, it might be assumed, therefore, that Cu(II) is bound to the same ligand atoms. Consequently, Cu(II) ions migth induce changes in the steric configuration of NAD.

This work was supported in part by Euratom grant EUR no. 213-76-7 BIO D.

EXPERIMENTS ON THE REACTION KINETICS BETWEEN "PLATINUM BLUE" COMPOUNDS AND SEVERAL BIOMOLECULES

P.Zaplatynski, H.Neubacher, W.Lohmann

Institut für Biophysik, Strahlenzentrum der Justus-Liebig-Universität
D-6300 Lahn-Giessen

The antitumor activity of several platinum complexes derived from cis-$Pt(NH_3)_2Cl_2$ (PDD) has been well established and a few clinical applications have been reported thus far. Among these are "platinum blue" compounds, which are less toxic than PDD. The molecular mechanism of action is still unknown. Three properties of these colored platinum compounds have been investigated:

A) Extreme line broadening of the ^1H-NMR lines of the ligands

B) Paramagnetism, resulting from partially oxidized platinum ions, located in a polymeric structure

C) Absorption bands in the visible region of the optical spectrum

The optical absorption bands and the paramagnetism can be interpreted as properties of columnar stacks of platinum complexes. The interaction of these complexes with a few biomolecules(e.g. thymine, cytosine, cytidine, ascorbic acid) has been investigated in aqueous solution. The following results have been obtained:

1) Sharp NMR lines superimposed on a broad spectrum

2) Changes in the ESR spectra

3) Decrease of the optical absorption bands

The results show, that colored platinum compounds will be modified in the presence of the biomolecules used. The original structure must be decomposed by the reaction resulting in small diamagnetic Pt-species as well as smaller paramagnetic polymers. The decomposition of the polymer structures of the colored platinum complexes is suggested to be due to biomolecules, which act as electron donors. The results reported here seem to support the hypotheses, that colored platinum compounds may act as a pool for smaller Pt-species in biological systems.

This work was supported in part by Euratom grant EUR no. 213-76-7 BIO D.

INVESTIGATIONS ON THE MOLECULAR INTERACTION OF THE ANTICARCINOGENIC COMPOUND CIS-DICHLORODIAMMINEPLATINUM(II): THE FORMATION OF COLORED PLATINUM COMPLEXES

H.Neubacher, P.Zaplatynski, W.Lohmann
Institut für Biophysik, Strahlenzentrum der Justus-Liebig-Universität
Giessen

After the discovery of the antitumor activity of cis-dichlorodiammine platinum(II) (PDD), numerous blue colored complexes have been observed on reaction of the hydrolysis products of cis-PDD with various bio-molecules. Some of these complexes seem to be better antitumor drugs with diminished nephrotoxicity than the original cis-PDD compound. The structure of these colored platinum complexes is, however, still unknown.

All colored complexes investigated thus far in regard to their formation and their molecular behavior are paramagnetic and can be detected by means of electron spin resonance(ESR) spectroscopy.

The ESR as well as the optical results can be explained by assuming partially oxidized polymeric structures. The paramagnetism of most of the complexes investigated is caused, very likely, by Pt in its low spin d^7 configuration, Pt(III), with the free spin located in the d_z2-orbital, as can be concluded from the g-values. The structure of the spectra can be attributed to the hyperfine interaction with adjacent ^{195}Pt nuclei.

While the g-values of the colored Pt complexes are similar for certain classes of species, the optical spectra seem to depend on the ligand used. There is a great influence of the ligands on the energy and intensity of the optical absorption bands in the visible wavelength range. These bands seem to be due to $d_z2 \rightarrow p_z$ band transitions together with metal-to-ligand charge transfer.

In a few complexes investigated the ESR spectra are nearly identical in both the liquid solution and in the solid state. The optical absorption bands in solution and the corresponding bands in the photo-acoustic spectrum of the solid appear at the same transition energies. It has to be assumed, therefore, that at least some of the colored platinum complexes are present as polymers also in liquid solution.

This work was supported in part by Euratom grant EUR no. 213-76-7 BIO D.

MODIFICATION OF THE RADIATION EFFECT ON NUCLEOBASES
AND NUCLEOSIDES BY NITROIMIDAZOLE-DERIVATIVES
AND ASCORBIC ACID

A. Bahr, V. Penka and W. Lohmann

Institut für Biophysik

Justus-Liebig-Universität Giessen

Strahlenzentrum, Leihgesterner Weg 217

6300 Giessen, FRG.

Derivatives of nitroimidazoles are used succesfully as radiation sensitizers in clinical application, animal experiments, and
cell cultures. Their molecular mechanism of action is, however, still
unknown. Since the nucleic acids seem to be the primary target of
radiation, the effect of some nitroimidazole-derivatives (Ro 07-0582,
Ro 05-9963, Ro 07-1051, Ro 07-0207, Ro 11-3696) on nucleobases
(adenine, uracil, cytosine) and nucleosides (adenosine, uridine,
cytidine) irradiated with ^{60}Co-γ-rays was investigated spectro-
photometrically. Aqueous solutions of each entity and of these
mixtures using definite molar ratios were irradiated in air and
nitrogen atmosphere as well. In all cases studied a linear relationship between extinction and radiation dose was obtained. The OER
varied between 1.1 and 2.0. Contrary to in-vivo experiments, all
Ro-compounds used exerted a protective effect on the nucleic acid
constituents. When ascorbic acid was used instead of the Ro-compounds, a protective effect was also observed when the radiation
effect was studied immediately after preparing the solutions. In the
case of solutions which were stored 3 to 4 days at room temperature
and in the dark, vitamin C exerted a strong sensitizing effect. The
effect observed didn't depend on the molar ratio of the components
within the concentration investigated.

This work was supported in part by Euratom grant EUR
no. 213-76-7 BIO D.

RADIKALERZEUGUNG   IN   EINKRISTALLEN   DECARBOXYLIERTER

AROMATISCHER   AMINOSÄUREN

W. Karsten und A. Müller

Institut für Biophysik und Physikalische Biochemie

Universität Regensburg, Postfach 397, D-8400 Regensburg

Zusätzlich zu früheren Ergebnissen an Tryptamin·HCl (Try) und
Tyramin·HCl (Tyr) wurde jetzt die Radikalbildung in Einkristallen der decarboxylierten Aminosäuren Phenyläthylamin·HCl
(Phe), und Histamin·2HCl (His) durch Röntgenstrahlung zwischen
77 K und 300 K mittels ESR-Spektroskopie untersucht. Bei den
drei Substanzen Tyr, Phe und His trat im Gegensatz zu Tryptamin·
HCl Dissoziation eines H-Atoms von C_β auf, in keinem Fall je-
doch Desaminierung an C_α wie bei den entsprechenden Aminosäuren. In diesen drei Substanzen liegt eine gestreckte Kon-
formation der aliphatischen Kette vor, während sie in Try ge-
faltet ist. Bei Tyr war die isotrope Aufspaltung der beiden
C-Protonen dieses Radikals ausgeprägt und reversibel temperatur-
abhängig. Wie erwartet, machte sich die fehlende Carboxylgruppe
durch verstärkte Radikalbildung in den aromatischen Gruppen
bemerkbar. In Phe und Tyr wurde je ein Dissoziationsradikal
am Ring sowie im Phe, Tyr und His ein Assoziationsradikal ge-
funden (in Try zwei verschiedene).

Unerwartet waren die lichtinduzierten Reaktionen der strahlen-
erzeugten Radikale, die bisher nicht an aromatischen Aminosäuren
beobachtet wurden. Hier ist eine Art Kreisprozess zwischen den
beiden Assoziationsradikalen von Try besonders bemerkenswert.
Sie können nämlich reversibel ineinander umgewandelt werden und
zwar in einer Richtung durch Licht und in der Gegenrichtung
durch Wärme.

Section B

CELL- AND MEMBRANE RESEARCH

POLYMORPHISM OF PURE AND MIXED PHOSPHOLIPID MONOLAYERS

O.Albrecht, H.Gruler, E.Sackmann

Universität Ulm
Abt. für Exp. Physik III
D-7900 Ulm/Donau
Germany

With direct measured isobars in addition to the conventional isotherms it is possible to show, that very pure phospholipid-mono-
layers exhibit not only the well known main-transition and the
gaseous-liquid transition at very low pressures (both of first
order), but two additional transitions of second order. The phase-
diagrams were interpreted with the help of the Landau theory of
phase transitions (Journal de Physique, 39 (1978) 301).

For Cholesterol-DPPC mixtures the isobares give strong evidence
for the existence of a 1:2 complex.

The isotherms of DPPC-DMPA mixtures show clearly the existence of
a eutectic-like phase at about 50 mol%. This is shown on the
following figure.

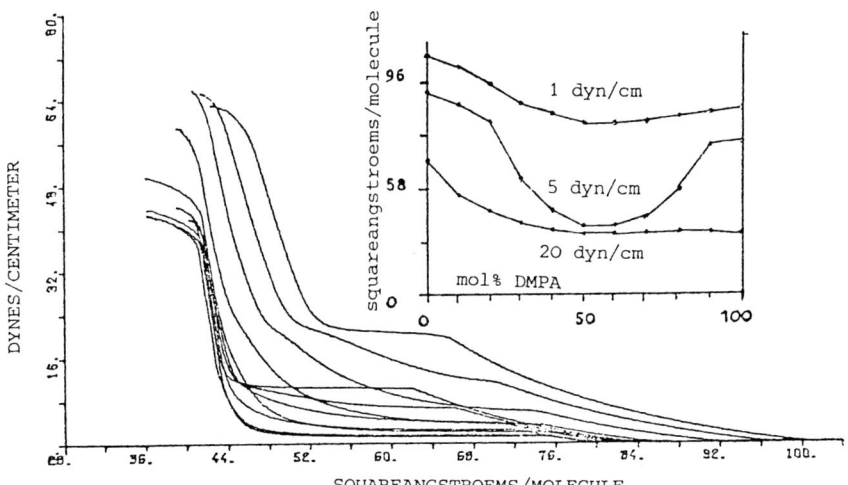

Isotherms of DPPC-DMPA mixtures 0-100 mol%, 29.9 deg Celsius on pure water. The
concentrations can be assigned with the help of the mixing-diagram in the insert

SPECTROSCOPIC INVESTIGATIONS OF MACROSCOPICALLY ORDERED
LECITHIN - WATER - MULTIBILAYERS

E. Jürgens, E. Sackmann
Universität Ulm
Abteilung Experimentelle Physik III
Oberer Eselsberg
D-7900 Ulm (Donau)

Macroscopically ordered multibilayers of dipalmitoyl phosphatidylcholine (DPPC) and dioleoyl phosphatidylcholine (DOPC) with a thickness of up to 2oo μm are prepared by a temper process. The water
content is varied up to 2o weight % by controlling the external humi-
dity and is measured by infrared spectroscopy.

EPR spin label spectroscopy using nitroxide fatty acid labels as
well as nitroxide lecithin labels gives a phase diagram of these thermotropic and lyotropic smectic liquid crystals. Especially the coexistence region of the fluid L_α phase and the crystalline $L_{\beta'}$ phase
(Luzzati notation) is investigated. The tilt angle of the molecules
with respect to the bilayer normal is determined using the anisotropy
of the hyperfine splitting of the spin label molecule. For DPPC with
1o weight % of water the tilt angle at 25 ^{O}C is about 4o ^{O}. It remains
fairly constant up to the beginning of the coexistence region (5o ^{O}C -
55 ^{O}C), where it abruptly drops down.

Preliminary Raman spectroscopic investigations using the intensity
ratios of the C-H-stretching bonds confirm the EPR results.

Freeze fracture electron micrographs give information about the
coexistence of water-rich and water-depleted areas.

The results are discussed together with literature data from small
angle X-ray scattering and from optical birefringence measurements,
which differ in the temperature of the coexistence region.

NEUTRON SMALL ANGLE SCATTERING ON MODELL MEMBRANES

W. Knoll[1], E. Sackmann[1], and H.B. Stuhrmann[2]

[1]Experimentelle Physik III, University of Ulm

[2]E.M.B.L., c/o DESY, Hamburg

Small angle neutron scattering experiments have been performed with dilute dispersions of lipids and lipid mixtures in order to study the lateral organization of model membranes. It is shown that this technique, being well established for the study of biopolymers in solution, also has several advantages for the investigation of lipid systems and even unique possibilities when compared with other methods:

i) During the experiment the sample remains in its thermodynamical equilibrium thus preventing artefacts arising from kinetic effects (as in differential scanning calorimetry) or from special preparation methods (as in electron microscopy).

ii) Deuteration of one of the components in a lipid alloy, which is a modification of minor influence on the physical properties of the molecules, yields a large contrast for neutrons relative to the other protonated species.

iii) Scattering experiments in different H_2O/D_2O ratios, i.e. giving different contrasts of the solvent relative to the sample, allow one to distinguish between homogeneous lipid mixtures and phase separation phenomena.

Some of the presented results are the following:

i) Dimyristoyl- (DML) and dipalmitoyllecithin (DPL), two lipids that differ only by two CH_2 groups in their hydrocarbon chains, are homogeneously distributed in a mixture in the ordered as well as in the fluid state. From the mean scattering length density, a positive excess volume of 107 $Å^3$ at 5°C and 57 $Å^3$ at 35°C being below and above the chain melting, respectively, is deduced. This correlates fairly well with the free volume that is created by a DML-molecule surrounded by the longer DPL-molecules.

ii) For a mixture of dimyristoyl- and distearoyllecithin, differing by four CH_2 groups, a phase separation in the ordered state could be established. Assuming one phase to be pure DML, the other can be calculated to contain 57% DSL with a positive excess volume of about 250 $Å^3$. Moreover, it could be demonstrated that these lipids do not mix homogeneously even in the fluid state.

iii) For another fluid mixture, dimyristoyllecithin and dimyristoylphosphatidic acid at 50°C, a similar domain formation could be shown. For this lipid alloy, however, a negative excess volume of about 12% is deduced.

KINETIC INVESTIGATIONS OF THE PHASE TRANSITION OF LECITHIN BILAYERS

Alfred Blume[1], Bernd Gruenewald[2] and Fumiyuki Watanabe[2]

1) Institut für physikalische Chemie II der Universität Freiburg,
   Albertstraße 23a, D-7800 Freiburg, West-Germany
2) Biozentrum der Universität Basel, Klingelberstrasse 70,
   CH-4056 Basel, Switzerland

Pressure jump experiments were performed on vesicles and liposomes of
dimyristoyl-phosphatidylcholine (DMPC) and dipalmitoyl-phosphatidyl-
choline (DPPC) following the time course of the solution turbidity in
the temperature range of the transition. Pressure jumps of 55 bar for
vesicles and 150 bar for liposomes were applied. It could be shown
that vesicles are not broken up to a notable extent when small pressure
jumps are used. The two relaxation times evaluated for both lipids
exhibit clear maxima at the midpoint of the transition. The time constants lie for DMPC vesicles in the 100 μs and 1 ms range and for DMPC
liposomes in the 1 ms and 10 ms time range. The relaxation processes
observed for DPPC vesicles and liposomes are slightly slower than those
for DMPC. For DPPC a third relaxation time in the 100 μs range can be
resolved at the midpoint of the transition. From the total pressure
jump amplitudes we can conclude that a fast process with small amplitude occurs outside the transition region and in the case of liposomes
also within the transition region. All relaxation times observed are
concentration independent. The time constant and amplitude behaviour
with maxima at the transition midpoint indicate that all processes are
cooperative in accord with previous interpretations. The integrated
p-jump amplitudes for vesicles result in transition curves in good
agreement with those obtained from static experiments. It is demonstrated that cooperative units can be evaluated from the relaxation amplitudes for vesicle solutions. These are of the same order of magnitude
as those calculated from static experiments. On the grounds of the
present kinetic investigations we can state that the application of
the linear Ising model to two-dimensional processes as attempted for
the static lipid phase transition is inadequate.

THE INTERACTION BETWEEN L-ASCORBIC ACID, α-TOCOPHEROL,
AND PHOSPHOLIPID MODEL MEMBRANES

A.Pleyer-Weber, H.Sapper, W.Strobelt, W.Lohmann

Institut für Biophysik, Strahlenzentrum der Universität,
Leihgesterner Weg 217, D-6300 Gießen

Recently a synergistic redox-interaction of α-tocopherol
(vitamin E) and ascorbic acid (vitamin C) has been proposed
concerning the lipid-antioxidant mechanism of these two vitamins[1]
A basic requirement for this mechanism should be an appropriate
molecular contact between α-tocopherol, ascorbic acid, and
membrane lipids.

In order to elucidate this mechanism the interaction of ascorbic
acid, α-tocopherol, and phospholipid model membranes
(L-ß,γ-dipalmitoyl-α-lecithin and L-ß,γ-dimyristoyl-α-lecithin
vesicles) were studied by means of ESR and NMR spectroscopy as well
as electron microscopy.

Vitamin E was found integrated in the lipid region of the membranes
due to the hydrophobic interaction of its phytyl side chain and
coordinated additionally to the polar head groups of the lipids.
Vitamin C was found associated in the polar region of the membranes
but penetrating also to the lipid region to some extent. Thus, the
interaction between the two vitamins seems to be facilitated by
the amphiphilic nature of the lipids and mediated by their polar
groups. There are some indications that the redox-interaction of
vitamin E and vitamin C is also supported by certain divalent
metal ions.

1) Packer,J.E.,Slater,T.F.,Willson,R.L.:Nature 278(1979)737-738

THE INTERACTION OF SYNTHETIC PHOSPHOLIPIDS WITH MELITTIN AS A MODELL TO STUDY LIPID - PROTEIN INTERACTIONS

H.Vogel, Max-Planck-Institut für Biologie, D-7400 Tübingen

The polypeptide melittin is composed of a mainly hydrophobic core of 20 aminoacids and a hydrophilic part of 6 aminoacids at the carboxylic terminus, which carries 4 positive charges at pH 7. Two different membrane systems are used - dimyristoyllecithin (DML) and the negatively charged dimyristoylmethylphosphatidic acid (DMMPA) - in order to study i) the structural changes of melittin upon binding to membranes and ii) the influence of the polypeptide on the physical properties of the lipid membranes.

Measurements of the circulardichroism (CD) indicate that melittin assume an unordered structure in aqueous solution, whereas in the membrane-bound state about 70% of the polypeptide chain consists of helical regions. This conformation is the same in the ordered and the fluid state of the membrane and is independent of the nature of the lipid headgroup. In DML membranes, the number of lipid molecules which constitute the binding site of one melittin molecule are 96 below the phase transition temperature T_t and 59 above T_t. In negatively charged membranes like DMMPA, only 4 lipid molecules are bound to one melittin molecule. The binding constants evaluated from CD-titration curves increase in the order DML ($T<T_t$), DML ($T>T_t$), DMMPA ($T<T_t$).

From quenching studies of the intrinsic fluorescence, the tryptophyl residue of melittin is located in the lipid headgroup region of the membranes

As seen by Raman-Spectroscopy, the binding of melittin to DML membranes decrease the concentration of trans conformers in the lipid chains below T_t and increase it above T_t. The width of the phase transition of DML-melittin complexes is greater than in pure DML membranes with the same T_t. X-ray scattering of this membrane preparations show that the bilayer structure of the membranes is still present. In DMMPA-melittin complexes a new phase transition is observed 20 $^{\circ}$C below the T_t of pure DMMPA membranes.

EINFLUSS KÜNSTLICHER IONOPHORE AUF DIE STRUKTUR

VON MODELLMEMBRANEN

B.Tümmler, U.Hermann, G.Maaß, F.Vögtle[+]

Das fluoreszierende 9,10-Dimethylanthracen (9,10-DMA) gehört zu einer Reihe von Fluorophoren, die nach Inkorporation in Membranen Sonden für den Phasenübergang in Lipiden vom kristallinen in den flüssig-kristallinen Zustand darstellen. Die Kopplung von zwei in 9,10-Stellung substituierten Anthracenmolekülen mit einem oder zwei Diaminopolyetherringen führt zu Komplexonen, die Metallionen spezifisch binden und gleichzeitig als Sonden für die Polarität und Beweglichkeit ihrer Umgebung fungieren können. An diesen Liganden werden die spezifischen Wechselwirkungen von Ionentransportsystemen mit den unmittelbar benachbarten Lipidmolekülen exemplarisch untersucht.
Die Liganden wurden in multi- oder bilamellare Phospholipidvesikel inkorporiert. In Abkühlkurven wurde die Änderung der Fluoreszenz in Abhängigkeit von der Temperatur registriert. Variiert wurden Salz und Ionenstärke.
Ergebnisse: 1. Die Liganden immobilisieren die umgebende Lipidmatrix.
2. In den differentiellen Abkühlkurven wurde neben dem Vor- und Hauptübergang der Phospholipide noch eine weitere Strukturänderung der Lipidmatrix beobachtet, die zur Zeit thermodynamisch und kinetisch charakterisiert wird.
3. Abhängig von der Membranfluidität befinden sich die Komplexone an Orten unterschiedlicher Polarität.

Institut für Klinische Biochemie II und Biophysikalische Meßgeräteabteilung, Medizinische Hochschule Hannover, 3000 Hannover 61
+ Institut für Organische Chemie und Biochemie der Universität Bonn, 5300 Bonn 1

^{31}P-NMR STUDIES ON THE LIPID CONTAINING BACTERIOPHAGE PM2

Hideo Akutsu, Haruhiko Satake, Richard M. Franklin and Joachim Seelig
Biocentre of the University of Basel

The lipid containing bacteriophage PM2 grows on the marine bacterium
Alteromonas espefiana. In the virus, phosphatidylglycerol (PG) and phospha
tidylethanolamine (PE) are the major components of the outer- and the
inner leaflets of the lipid bilayer, respectively . Originally it was
reported that the nucleocapsid of this virus, which can be isolated in
the presence of 4 M-8 M urea, has no lipids (1). In order to correlate
the viral structure with the biological function, phosphorus nuclear
magnetic resonance (^{31}P-NMR) studies on PM2 have been carried out. A
powder pattern spectrum of PM2 solution was successfully obtained in
the presence of either 50 or 60% sucrose by ^{31}P-NMR at 36.43 MHz. The
spectrum is composed of two major components. One is a pattern typical
of an axially symmetrical motion and this was assigned to the phospholipid bilayer of PM2 by comparison with the spectrum of the extracted
lipid from PM2 and reported data by other methods. The chemical shift
anisotropy was about 47 ppm at 6°C. The other component was much broa-
der. This was assigned to the packaged DNA of PM2 by comparison with
the ^{31}P-NMR spectrum of bacteriophage T_4, a virus which has no lipids.
A powder pattern spectrum of the PM2 nucleocapsid was also successfully
obtained in the presence of 6 M urea and 50% sucrose. The spectrum was
quite similar to that of PM2. This fact clearly shows that the nucleocapsid still contains a phospholipid bilayer in contradiction with the
proposed model, and that the structural arrangement of the lipid bilayer and DNA is similar to that of the intact PM2. The existence of
phospholipid in the nucleocapsid was also confirmed by biochemical
analysis.

In the temperature shift experiments a spectral change was observed at
about 17°C both for the phospholipid and DNA components. The purified
PE vesicles showed a phase transition temperature close to 17°C, while
PG vesicles had no observable transition between 0°C and 35°C. Furthermore the effect of temperature on the infectivity of this virus in the
early stage of the growth cycle was examined. The efficiency of the
infection increased in a step-wise fashion in the range from 15°C to
20°C. These facts suggest a correlation between the structural change
of the virus and its infectivity. (1) R. Schäfer, P. Künzler, A. Lustig
and R.M. Franklin (1978) Eur.J.Biochem. 92, 579-588.

THE HEAD GROUP MOBILITY IN PHOSPHATIDYLCHOLINE MEMBRANES IN THE PRESENCE OF CHOLESTEROL AS SEEN BY A DIELECTRIC INVESTIGATION

John C. W. Shepherd and Georg Büldt
Department of Biophysical Chemistry, Biocentre of the University of Basel,
Klingelbergstrasse 70, CH-4056 Basel

For phospholipid membranes with zwitterionic head groups, the dipole can be considered as a specific label for tracing the changes in the dynamic behaviour of this region of the bilayer in its various phases. Measurements of the dielectric properties of fully hydrated 1,2-dipalmitoyl-sn-glycero-3-phosphocholine bilayers in the frequency range 1 - 50 MHz show a dispersion which is attributed to the motion of the phosphocholine dipoles in the plane of the bilayers. When the temperature is varied, both the permittivity and loss factor increase sharply at the pretransition $(35^\circ C)$ and the main transition $(42^\circ C)$. The relaxation time and amplitude were also determined for this dispersion and these further reflect the structural changes occurring with temperature. Due to steric hindrances a restriction in the angle of head group rotation occurs at lower temperatures but is greatly reduced above the main transition. When cholesterol is added the abrupt change in the derived relaxation frequency f_2 observed for pure DPPC at the gel-to-liquid crystalline phase transition at $42^\circ C$ reduces to a more gradual increase of frequency with temperature as the cholesterol content is increased. In general the presence of cholesterol increases the DPPC head group mobility due to its spacing effect. Below $42^\circ C$ no sudden changes in f_2 are found at 20 or 33 mole $^\circ/o$ cholesterol, where phase boundaries have been suggested from other methods. Above $42^\circ C$, however, an initial decrease in f_2 at cholesterol content up to 20 - 30 mole $^\circ/o$ is found. This is thought to be partly due to an additional restricting effect of the cholesterol on the number of hydrocarbon chain conformations and consequently on the area occupied by the DPPC molecules.

A TIME AVERAGED PICTURE OF THE CONFORMATION AND SEGMENTAL DISORDER OF PHOSPHATIDYLCHOLINE IN BILAYERS INVESTIGATED BY NEUTRON DIFFRACTION

G. Büldt, H. U. Gally, A. Seelig, J. Seelig, and G. Zaccai[*]
Department of Biophysical Chemistry, Biocentre of the University of Basel, Switzerland, and [*]Institut Laue-Langevin, Grenoble, France

Neutron diffraction experiments on selectively deuterated lipids provide a new method of determining to a segmental resolution the mean conformation of a lipid molecule as projected along the bilayer normal, despite the high amount of disorder that exists in these bilayers. In addition a time averaged picture of the extent of the positional fluctuations of the individual segments in this direction can be given. This is demonstrated for a multilamellar system of bilayers of 1, 2-dipalmitoyl-sn-glycero-3-phosphocholine (DPPC). Samples deuterated at 13 different positions were measured as oriented samples and as unoriented samples in the gel phase L_β, and in the liquid crystalline phase L_α. The derived structure factors for the deuterated segments were fitted assuming a Gaussian distribution of the segments along the bilayer normal, from which the mean label positions were determined with a precision of ± 1 Å in most cases. The results clearly show that the average orientation of the zwitterionic phosphocholine group is almost parallel to the membrane surface in the gel state ($L_{\beta'}$) as well as in the liquid crystalline state (L_α). For the hydrocarbon region the values obtained in the $L_{\beta'}$ phase confirm the model with chains in all trans configuration tilted with respect to the bilayer normal by an angle which increases with water content. From samples that were deuterated in both chains separately and studied at low water content it was seen that the chains of the DPPC molecule are out of step by as much as 1.5 carbon-carbon bond lengths. A constant width of the label distribution in the projection on the bilayer normal was observed for segments at the beginning and end of the chains. This is an additional indication for the chains being in the all trans state in $L_{\beta'}$ phase. In the L_α phase, the present experiments show that consecutive segments are well separated in the profile. The whole chain region is shortened by a factor of ~ 0.75 compared to the $L_{\beta'}$ phase. In contrast to the gel phase the width of the label distribution is not constant over the entire region, but is found to be increased by more than a factor of two at the end of the chains.

DETERMINATION OF LIPID ORDER PARAMETERS FROM FLUORESCENCE DEPOLARIZATION EXPERIMENTS.

M. P. Heyn

Dept. of Biophysical Chemistry, Biozentrum, CH-4056 Basel, Switzerland.

The equilibrium orientational distribution function $f(\theta)$ of a rigid fluorescent probe embedded in a membrane can be expanded in the complete set of orthogonal Legendre polynomials:

$$f(\theta) = \sum_{\ell = 0}^{\infty} c_\ell \, P_\ell \, (\cos\theta), \text{ with } c_\ell = \frac{2\ell + 1}{2} \int_0^\pi P_\ell \, (\cos\theta) \, f(\theta) \, d\cos\theta$$

θ is the angle between the long axis of the probe and the normal to the membrane. $f(\theta) \, d\theta$ is the probability that the probe has an orientation between θ and $\theta + d\theta$. For odd ℓ, c_ℓ is zero. The moments c_ℓ of the unknown distribution function thus characterize $f(\theta)$ completely. Provided a sufficient number of c_ℓ's can be determined experimentally and provided the series converges rapidly, the truncated series will give a good representation of $f(\theta)$. The familiar order parameter S is just $2/5 \, c_2$. In recent fluorescence depolarization experiments on rigid-rod like probes embedded in lipid bilayers it was shown that the fluorescence anisotropy $r(t)$ does not decay to zero at long times, but rather reaches the constant value r_∞ indicating that the equilibrium orientational distribution function is anisotropic. The theoretical approach of Kinosita (1) can be used to show that for probes with parallel absorption and emission transition dipole moments, the initial (r_0) and final (r_∞) values of the anisotropy are related to the order parameter S in the following way: $(r_\infty/r_0) = S^2$. Thus the order parameter of a small fluorescent lipid probe and by assumption of that of the acyl chains of neighbouring lipids, can be determined from the experimentally accessible quantities r_0 and r_∞. Moreover, c_4 can be determined from fluorescence depolarization experiments with oriented membranes (2). A systematic approach for obtaining $f(\theta)$ by the measurement of its moments is thus available. The present approach will be compared with the "cone model" in which an unrealistic form of $f(\theta)$ is imposed on the data (1). A comparison between order parameters from fluorescence depolarization and deuterium NMR will be presented. Cholesterol and proteins appear to increase the order parameter of fluorescent lipid probes.

1. Kinosita, K., Jr., Kawato, S., and Ikegami, A. (1977) Biophys. J. 20, 289.

2. Chapoy, L. L., and DuPre, D. B. (1979) J. Chem. Phys. 70, 2550.

LIPID-PROTEIN INTERACTIONS IN BACTERIORHODOPSIN-PHOSPHATIDYLCHOLINE VESICLES.

M. P. Heyn[+], N. A. Dencher[+], and R. J. Cherry[*]
[+] Dept. of Biophysical Chemistry, Biozentrum, 4056 Basel, Switzerland,
[*] Dept. of Biochemistry, ETH Zürich, 8092 Zürich, Switzerland.

Of considerable current interest is the question how the presence of proteins in membranes affects the order and mobility of the membrane lipids. In an attempt to understand the effect of proteins on the structural and dynamical parameters of lipids it is of great advantage to be able to vary the lipid to protein ratio (L/BR). In the present investigation we use DMPC/bacteriorhodopsin vesicles in which a lowering of the temperature from above the gel to liquid-crystalline transition of the lipids to below induces a reversible change in the aggregation of BR from a monomeric state above to the crystalline state below T_c. In these vesicles the L/BR can be varied from 18 to 390 (molar), the photocycle is slowed-down but otherwise normal and BR is functional as a light-driven proton pump both above and below T_c (Dencher and Heyn, in press). The lipid phase transition was monitored using the steady-state anisotropy of the fluorescent probe diphenylhexatriene (DPH). Below T_c the anisotropy is quite high (0.34) and independent of the L/BR. The lipid phase transition, which is accompanied by a large drop in anisotropy, broadens with decreasing L/BR. Within experimental error T_c is the same as for protein free vesicles. Above T_c the anisotropy increases markedly with decreasing L/BR. This effect can be explained either by assuming that BR slows down the rotational diffusion of the lipids (viscosity) or by assuming that BR restricts the angular range available for rotation (order). It can be estimated that the observed effect cannot be due to a protein-induced increase in the viscosity alone. Results of time-dependent fluorescent depolarization experiments on other systems also show that the probe rotational correlation time increases only slightly when proteins are incorporated in lipid bilayers. We are thus led to the conclusion that addition of BR increases the lipid order parameter. Changes in the dynamics and in the aggregation state of BR were monitored by measurements of its rotational diffusion constant and of its CD spectrum. In the former method the immobilization of BR which accompanies the formation of the crystalline protein patch is followed by the changes in its rotational diffusion constant. In the latter method the formation of specific BR aggregates is monitored using the exciton bands in the CD spectrum which are due to interactions between retinal chromophores on adjacent BR's. BR crystallizes several degrees below the lipid phase transition. The effect of the L/BR ratio on ΔH and T_c of the transition will be discussed. Above T_c the membrane viscosity as determined from rotational diffusion measurements is 2 to 3 times larger as that obtained from fluorescence depolarization measurements.

AGGREGATED AND MONOMERIC BACTERIORHODOPSIN PUMPS PROTONS

Norbert A. Dencher and Maarten P. Heyn

Biozentrum, University of Basel, CH-4056 Basel, Switzerland

Bacteriorhodopsin (BR) molecules are arranged as immobilized trimers in the hexagonal crystalline lattice of the purple membrane. Experiments were designed to answer the question whether the hexagonal state of aggregation is required for proper functioning of this light-driven proton pump. Using the detergents Triton X-100 and octylglucoside BR was solubilized and incorporated as monomers into large unilamellar lipid vesicles. The thermotropic behaviour of these vesicles, i.e. of the lipid phase and of the protein-protein interactions, was monitored by means of the fluorescence polarization probe DPH and circular dichroism (CD) measurements, respectively. Light-induced pH-changes of vesicle suspensions were observed with a pH-electrode. The presence of exciton coupling features in the visible CD spectrum of aggregated BR and its absence in monomeric BR allow the determination of its aggregation state. In accordance with previous work[1] the results indicate that at temperatures below the lipid phase transition, BR crystallizes into patches with the same hexagonal lattice as is observed in purple membrane, whereas above the transition the lattice disaggregates and the BR molecules are monomeric. In DMPC-BR vesicles the CD transition curve caused by changes in the aggregation state of BR is shifted by about 6-8°C towards lower temperatures in comparison to the lipid phase transition. The midpoint of the latter transition (about 23°C) was not changed due to the presence of BR. Illumination resulted in an alkalization of the external medium independently of the BR aggregation state. The steady-state values of the pH-changes at 10°C (aggregated BR) and 30°C (monomeric BR) represent 1.6 and 3 H^{+}/BR translocated, respectively. In egg lecithin and asolectin vesicles at 20°C, i.e. far above the lipid phase transition, net inward pumping of up to 12 H^{+}/BR was observed. The alkalization was found to be inhibited by proton ionophores indicating an H^{+}-transport across the membrane. Whereas the channel gramicidin abolished the H^{+}-gradient at all temperatures tested, the H^{+}-carrier CCCP functioned only above the phase transition. Our experiments show that the proton pumping capacity of BR is independent of its state of aggregation. This seems to be true for the photochemical cycle as well (K.D. Kohl, W. Sperling, N.A.D. and M.P.H., unpublished results). Whereas assembly into specific aggregates seems to be a necessary prerequisite for the function of other membrane transport proteins, monomeric BR itself can effectively pump protons.

1. R.J. Cherry, U. Müller, R. Henderson, M.P. Heyn. J. Mol. Biol. (1978) 121, 283.

ON THE INTERPRETATION OF LIPID MEMBRANE COOPERATIVITY DATA

Stefan Stankowski and Bernd Gruenewald
Biozentrum der Universität Basel, Switzerland

Information about the cooperative character of phospholipid phase transitions can be deduced from the evaluation of transition curves. Especially for temperature-induced transitions of liposomes and vesicles a considerable amount of data has been accumulated. Using calorimetrically determined enthalpy changes, so-called "cooperative units" can be obtained from the midpoint slopes of the curves.

The correct interpretation of these data, however, has encountered difficulties, because the theoretical models discussed so far are mostly molecular-field models and include the cooperativity only in a very rough approximation. We propose therefore to apply the two-dimensional Ising model in the quasi-chemical approximation which is both powerful and easy to handle. Explicit formulae for the evaluation of cooperative units and cooperativity parameters can then be given. The model allows a direct comparison with the well-known Ising results for linear chains. In addition, the question is discussed whether the finite slope of transition curves of liposomes is in contradiction with the assumption of a first-order character of phospholipid phase transitions.

The model can easily be extended to the treatment of lipid systems containing cholesterol, general anaesthetics or proteins. In all these cases, the formalism always remains simple and suitable for straightforward physical interpretation.

STATISTICAL DESCRIPTION OF PHOSPHOLIPID BILAYERS WITH APPLICATION TO DEUTERIUM NMR

J.-P. Meraldi[a], J. Schlitter[b], and J. Seelig[a], [a]Biocenter, University of Basel, CH-4056 Basel, Switzerland, and [b] Biophysics Department, University of Bochum, D-4630 Bochum, Germany.

In order to reproduce accurately and to extend the highly qualitative deuterium NMR data on the organization of acyl chains in phospholipid bilayers, we have developed a model which follows closely the generalized van der Waals approach recently applied to nematic liquid crystals. The repulsive forces are considered as arising from hard core exclusions, whereas the attractive forces originate from the interaction of each molecule with its neighbors through a mean field. The dependence of the energy of interaction on the relative chain conformations has been approximated through consideration of the overall "shape" of a chain conformation and its corresponding "mean orientation". In the practical computations only chain configurations which are entering the polar head regions are rejected. The model which exhibits a first order phase transition (or a continuous transition under certain circumstances) allows a very accurate reproduction of C-D order parameters along the acyl chain (order profile) as well as their temperature dependencies. Analysis of several different conditions shows that the order profile is most sensitive to the distribution of chain configurations. Additionally these calculations emphasize the danger of making any "a priori" choices as to which conformations are to be allowed. The model has also been used to compute the mean distances between the CH_2 groups and the results were correlated with data from neutron scattering. Furthermore, we have calculated the mean number of trans and gauche segments, the probabilities of kinks and for each CH_2 group the ordering tensor was exactly determined. The mean spatial occupancy of each segment and the overall chain has been also studied. Finally we have compared our results with those given by Marcelja theory, a model which has initiated the present one but which did not consider repulsive forces.

STRUCTURAL ORDER OF LIPIDS AND PROTEINS IN MEMBRANES.
NEW EVALUATION OF FLUORESCENCE ANISOTROPY DATA

Fritz Jähnig
Max-Planck-Institut für Biologie
Corrensstrasse 38, D-74 Tübingen

The limiting long-time value of fluorescence anisotropy in
membranes is correlated with the orientational order parameter determined by deuterium magnetic resonance. Existing
experimental results for diphenylhexatriene in lipid bilayers
are evaluated for the order parameter, and a comparison is
made with values from deuterium magnetic resonance. Steady-
state measurements of fluorescence anisotropy can provide
the order parameter in good approximation. Proteins in the
fluid-lipid phase increase the lipid order parameter so
determined. Upon comparison with deuterium magnetic resonance
results it is concluded that proteins impose a tilted direction
on the surrounding lipids (see Figure below). Order parameters of protein segments obtained from the limiting value
of protein fluorescence anisotropy are discussed as to the
influence of lipid order on protein order.

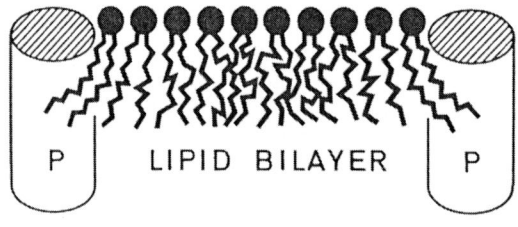

New Measurements of the Water-Permeability across
Lipid Bilayer Vesicles, the Effect of Cholesterol.

Rüdiger Lawaczeck
Institut für Physikalische Chemie der Universität
Marcusstr. 9/11, 8700 Würzburg, FRG

Indole chromophores fluoresce with different quantum yields, Φ, in D_2O and H_2O. Based on this solvent-isotope effect where $\Phi(D_2O) > \Phi(H_2O)$, a new technique to measure the permeation of water-molecules across vesicular lipid bilayers was developed (1). In principle an aqueous (H_2O) vesicle solution containing the encapsulated, water-soluble chromophore is rapidly mixed with the deuterated solvent. The permeation of the D_2O (or alternatively H_2O) molecules is monitored by the time-dependent increase of the fluorescence intensity. From this rise of the fluorescence, the exchange relaxation rate, k_{ex}, and, knowing the vesicular radius, the permeability coefficient, P_d, can be deduced. The experimental and theoretical background together with preliminary results will be published shortly (2).

Mixed vesicles containing increasing amounts of cholesterol embedded in a dipalmitoyl-lecithin matrix do not reveal the prephase transition phenomenon of the single component lecithin bilayers. Below the crystalline to liquid-crystalline phase transition temperature, T_c, the Arrhenius-type presentations of the data ($\ln(k_{ex})$ vs. $1/T$) are linear in contrast to cholesterol-free lecithin bilayers. An overall activation energy for k_{ex} of 21 to 23 kcal/mol is calculated for 5 to 20 mol% cholesterol in the temperature-range below T_c. At higher cholesterol to lipid ratios, i.e. 32 mol% cholesterol, preliminary results indicate a conspicuous decrease of the overall activation energy of k_{ex} to about 15 kcal/mol below the T_c of the pure dipalmitoyl-lecithin matrix. Similar but less pronounced effects around 33 mol% cholesterol have been observed in osmotic shrinkage experiments (3). Our early results seem to confirm numerous observations which show changes of bilayer properties starting around 30 mol% cholesterol incorporated into lecithin matrices. However, an interpretation in terms of a microscopic organization (complete miscibility, stoichiometric complexes and/or domain structures) of the cholesterol containing lipid bilayers is beyond the scope of the present experiments.

(1) R. Lawaczeck, J. Am. Chem. Soc. 100, 6521-6523 (1978)

(2) R. Lawaczeck, J. Membrane Biol., in press

(3) M. C. Blok, L. L. M. Van Deenen and J. DeGier, Biochim. Biophys. Acta 464, 509-518 (1977)

INVESTIGATIONS OF LIPOID PH INDICATORS AS PROBES FOR
ELECTROSTATIC POTENTIAL OF MICELLES AND VESICLES

A. Haase and P. Fromherz

MPI für Biophysikalische Chemie (Karl-Friedrich-
Bonhoeffer-Institut) D-3400 Göttingen, BRD

Measurements of the electrostatic surface potential of micellar systems with two fluorescent lipoid pH indicators, hydroxycoumarin and
aminocoumarin, has been described by M.S.Fernandez and P.Fromherz (1).
We show computer controlled titrations of these lipoid dyes, incorporated into ionic micelles (sodiumdecylsulfate, sodiumdodecylsulfate,
sodiumtetradecylsulfate, and hexadecyl-P-cholinester). With the experimental pK-shifts, the electrostatic potential and the dissociation
constant for these micellar systems are obtained. This procedure is
also applied for artificial lipid bilayers (phosphatidylcholine and
phosphatidic acid). These results are compared with two other lipoid
pH indicators (dimethylaminonaphthalin-sulfo-stearyl-amid and 3-pal-
mitoyl-7-oxy-coumarin).

For the determination of the electrostatic potential, it is important
to know the precise location of the chromophore within the micelle system and the lipid bilayer, which has been obtained with ^{1}H-NMR-measure-
ments. Micelles and lipid bilayers with incorporated chromophores show
upfield shifted ^{1}H-NMR lines since the aromatic ring gives a negative
contribution to the local magnetic field on the neighbor protons. This
effect is dependend on distance between the center of chromophore and
the neighbor protons and is only significant within a few angstroms.
From these measurements it is shown, that the two coumarin dyes are
located at the same region within the micelle or the lipid bilayer.
The hydroxy- and aminogroup of the indicator are located in the plane
of the sulfate groups for the micelles and the phosphate groups for
lecithin and phosphatidic acid.

(1) Fernandez,M.S., Fromherz,P., (1977) J.Phys.Chem. 81, 1755-1761

INFLUENCE OF CHOLESTEROL ON PERMEABILITY AND STRUCTURAL PROPERTIES
OF ARTIFICIAL MEMBRANES CONTAINING ENANTIOMERIC PHOSPHATIDYLCHOLINES
AND ETHER ANALOGUES.

A. Hermetter and F. Paltauf, Institut für Biochemie, Technische Universität Graz, A-8010 Graz.

The sn1- and sn3- isomers of dioleoylglycerophosphocholine (diester-PC)
form vesicles of the same size as the racemic lipid (r_o = 138 $\overset{o}{A}$, V_i =
= 0,45 µl/ µmol PC). Identical permeability coefficients were found for
glucose diffusion across membranes consisting of these lecithins. Racemic dioleylglycerophosphocholine (diether-PC) lacking two ester carbonyl
groups forms smaller vesicles (r_o = 122 $\overset{o}{A}$, V_i = 0,28 µl/ µmol PC), but
shows permeability coefficients for ions (Cl^-, Rb^+) and neutral molecules
(glucose) similar to those of diester-PC.
Cholesterol does not influence vesicle size up to a content of 30 mol %,
but reduces permeability of bilayers containing sn1-, sn3- and racemic
diester-PC as well racemic diether-PC to the same extent. Increasing
amounts of cholesterol (17, 33, 50 mol %) broaden the $(CH_2)_n$ signal in
the ^{1}H-NMR spectrum of unilamellar vesicles containing sn1-, sn3- and racemic dipalmitoylglycerophosphocholine (diester-PC) as well sn1-, sn3-
and racemic palmityloleoylglycerophosphocholine (ether-ester-PC) to a
similar degree indicating a decreased mobility of the hydrophobic residues in these phospholipid molecules.
The results show, that phospholipid-sterol interaction occurs mainly in
the hydrophobic region of the lipid molecules, and that no stereospeci-
fic requirements exist with regard to the phospholipid.

SURFACE POTENTIAL STUDIES OF MONOLAYERS FROM DIFFERENT STEROLS AND VA-
RIOUS PHOSPHOLIPID/STEROL MIXTURES.

P. Jauch, A. Hermetter[+], F. Paltauf[+] and R. Benz, Fachbereich Biologie,
Universität Konstanz, D-7750 Konstanz and [+]Institut für Biochemie,
Technische Universität Graz, A-8010 Graz

A surface potential instrumentation was constructed in order to measure
simultaneously surface potential as well as surface pressure of monolayers as a function of the area of the molecules. Monolayers from sterols with a 3-ß hydroxy group (cholesterol, 7-dehydrocholesterol, stigmasterol and ergosterol) showed potentials between 400 and 475 mV at
30 dyn/cm. Monolayers from epicholesterol (3-α hydroxy group) showed a
much smaller surface potential of 135 mV, possibly caused by the different orientation of the hydroxy group.
In the series cholesterol, 7-dehydrocholesterol, stigmasterol and ergosterol the "condensation" effect of monolayers from sterol/dioleoylle-
cithin mixtures decreased from 10 Å to about 2 Å at 30 dyn/cm. Conside-
ring the dipole moment of the single molecule in the mixed monolayers
defined as the potential per molecule, no dipole interaction could be
detected between dioleoyllecithin and cholesterol, whereas the dipole
moment is reduced in monolayers of ether-ester lecithin and cholesterol.
Optical active lipids such as d-lecithins and L-lecithins showed identical values for the surface potentials of monolayers, and no different
behavior in monolayers with different sterols could be detected for both
types of lipids.
A comparison between the surface potentials of lipid/sterol monolayers
and permeability studies with lipid bilayers from lipid/sterol mixtures
[1,2] shows that presumably only a small part of the sterol added to
the bulk phase is present in the black part of the membrane.

[1] Benz, R., Fröhlich, O., Läuger, P. Biochim. Biophys. Acta 464, 465
    (1976)
[2] Benz, R., Cros, D., Biochim. Biophys. Acta 506, 265 (1977)

INFLUENCE OF SURFACE POTENTIAL ON ION TRANSPORT THROUGH LIPID BILAYER
MEMBRANES

R. Schindler and R. Benz, Fachbereich Biologie, Universität Konstanz,
D-7750 Konstanz

Charge-pulse relaxation studies [1,2] were performed with different
transport systems in order to study the influence of surface charge and
surface potentials on ion transport through lipid bilayer membranes.
The membranes were made from a synthetic negatively charged phosphatidylserine (C16:CH$_3$-PS) and from a ether phosphatidylethanolamine
(C18:1-0-PE) which is neutral at pH 6 and negatively charged at pH 12.
Three different transport systems were used to study the influence of
the negative charges. In the case of the valinomycin-Rb$^+$-system the four
kinetic constants could be determined. It was found that the association rate constant k_R as well as the translocation rate constant k_{MS}
and k_S of the charged and the uncharged complex are strongly dependent
on the surface charge density of the membranes. Increasing the pH from
6 to 11.5 the deprotonation of the ammonium group in the C18:1-0-PE
leads to a 30 fold increase of k_R and a 5 fold increase of k_{MS} and k_S
which may be related to a fluidity change of the membranes which has
been measured in similar systems with other methods [3,4].
In experiments with dipicrylamine (DPA$^-$) and PV-K$^+$ it was found that
the partition coefficient of the lipophilic ion and the carrier-ion
complex is strongly dependent on the surface charge density. From experiments using different ionic strength in the presence of DPA$^-$ with
membranes from C16:4CH$_3$-PS a concentration of negative charges on the
membrane surface of 1 charge per 0.55 nm^2 could be measured. All experiments with negatively charged membranes could quantitatively be explained by the Gouy-Chapman theory and with surface potential studies
of monolayers.

[1] Benz, R., Läuger, P. J. Membrane Biol. <u>27</u>, 171 (1976)
[2] Benz, R., Läuger, P., Janko, K. Biochim. Biophys. Acta <u>455</u>, 701
    (1976)
[3] Träuble, H. Eibl, H. Proc. Natl. Acad. Science U.S.<u>71</u>, 214 (1974)
[4] Eibl, H., Blume, A. Biochim. Biophys. Acta <u>553</u>, 476 (1979)

INFLUENCE OF LOCAL ANESTHETICS ON ION TRANSPORT THROUGH LIPID BILAYER MEMBRANES

G. Ehmann and R. Benz, Fachbereich Biologie, Universität Konstanz, D-7750 Konstanz

Charge pulse experiments [1,2] were performed in order to study the influence of local anesthetics such as procaine, benzocaine, tetracaine, lidocaine and dibucaine on ion transport through lipid bilayer membranes. The neutral lipids dioleoyllecithin (18:1 PC), dioleoylphosphatidylethanolamine (18:1 PE) as well as the negatively charged brain-phosphatidylserine (PS) were used for membrane formation. In kinetic studies with dipicrylamine (DPA^-) and $PV-K^+$ the translocation rate constants k_i and k_{MS} for the lipophilic ion and the carrier ion complex could be measured in the presence of the local anesthetics. It was found that the partition coefficient ß of the negatively charged lipophilic ion is strongly dependent on the local anesthetics and decreases in the series procaine to dibucaine. This finding is in contrast to the hypothesis that the action of local anesthetics can be explained by an adsorption of positive charges to the membrane [3]. Our results suggest that besides an adsorption of positive charges (some local anesthetics are positively charged at pH 5-7) the dipole potential in lipid bilayer membrane (usually several hundred mV) decreases drastically by the adsorption of the local anesthetics. Its efficiency increases therefore with the increase of the hydrophobic part of the molecule.
Experiments with lipid monolayers in the presence of local anesthetic show that the surface potential is reduced by more than 100 mV. This finding is so far consistent with the above mentioned suggestion that local anesthetics reduce the dipolar potential of membranes.

[1]   Benz, R. Läuger, P., J. Membrane Biol. 27, 171 (1979)
[2]   Benz, R., Läuger, P., Janko, K., Biochim. Biophys. Acta 455, 701 (1979)
[3]   McLaughlin, S. in Progress in Anesthesiology, Vol. 1, Raven Press, N.Y. 1975, p. 193

NEGATIVE HYDROPHOBIC IONS AS CARRIERS FOR POSITIVE HYDROPHOBIC IONS

G. Stark

Fakultät für Biologie, Universität Konstanz, D-7750 Konstanz

The permeability of positively charged hydrophobic ions across biological membranes and artificial lipid membranes is strongly increased
in the presence of trace amounts of negative hydrophobic ions (1).
This effect is frequently used to accelerate the distribution of
hydrophobic cations such as tetraphenylarsonium ($TPAs^+$) across cell
membranes in order to estimate the magnitude of the electric membrane
potential.

The catalytic action of anions A^- like dipicrylamine (DPA^-) or tetra-
phenylborate (TPB^-) on the permeability of cations B^+ like $TPAs^+$ is
explained here on the basis of ion pair formation B^+A^-; i.e. the free
anions A^- are believed to function as carriers for the cations B^+.
The partition coefficient for A^- between membrane and water is usually considered to be much larger than that of the cations B^+. This is
because of the sign of the electric dipole potential at the membrane/
water interfaces. The concentration of B^+ inside the membrane would
be strongly increased by formation of electrically neutral ion pairs
B^+A^-.

The experimental evidence is based on voltage jump relaxation experiments. It was found that the concentration of the free anions A^- is
reduced as the concentration of the cations B^+ is increased in fair
agreement with the simple carrier model mentioned above. The dissociation constant K for ion pair formation in water between DPA^- and
$TPAs^+$ was estimated from the concentration dependence of the current
relaxation ($K = 2,5 \cdot 10^{-5}$ M).

The model could be of more general importance for the understanding
of a facilitated permeation of hydrophobic ionic substances across
cell membranes mediated by integral ionic membrane components of
opposite charge.

(1) Ye.A. Liberman and V.P. Topaly, Biofizika 14, 452 (1969)

LASER-TEMPERATURE-JUMP METHOD FOR THE DETERMINATION OF THE RATE OF DE-
SORPTION OF HYDROPHOBIC IONS IN LIPID MEMBRANES

W. Brock, P.C. Jordan[+] and G. Stark, Fakultät für Biologie, Universität
Konstanz, D-7750 Konstanz and [+]Department of Chemistry, Brandeis University, Waltham, M.A. 02154, U.S.A.

In a previous paper (1) a general treatment of the coupled diffusion
process for hydrophobic ions through both the aqueous phase and the membrane was presented. The system phosphatidylserine/tetraphenylborate of-
fered the most favourable conditions for estimating the rate constant k
of desorption. From voltage-jump experiments the authors concluded that
k was comparable to the rate of translocation.
Here T-jump experiments are used in order to test the theory and to determine k by an independent method. The T-jump was generated by a Nd-
glass-laser with a wavelength of 1.06 μm, a pulse energy of 17 Joule
and a pulse duration of 500 μs. The resulting T-jumps ranged from 0.4
to 0.8 K. The relaxation of the current carried by the hydrophobic ions
was monitored by a series of voltage jumps. Thus, the effect of diffusion polarization on the current measurement could be avoided. The theoretical curves for the current relaxation depend on three parameters:
the diffusion coefficient D, the adsorption partition coefficient ß and
k. For D a literature value was used, and ß was determined by the voltage jump method. Hence, only k had to be varied to fit the theoretical
curves to the experimental data.
Under the assumption that the adsorption/desorption process is voltage
independent, a good agreement between theoretical and experimental values could be obtained. The average value of k found for the best fits
is 60 s^{-1} and coincides with the value given in (1). From the experimental error limits k is found to lie in the interval 20 s^{-1}< k $\lesssim$ 600 s^{-1}.
Hence both limiting cases can be excluded:neither is diffusion fast nor
can the current relaxation be explained by diffusion alone.

(1) P.C. Jordan and G. Stark, Kinetics of transport of hydrophobic
    ions through lipid membranes including diffusion polarization in
    the aqueous phase, submitted.

VOLTAGE DEPENDENT NOISE CURRENT OF LIPID BILAYER MEMBRANES GENERATED BY HYDROPHOBIC IONS

R. Junges and H.-A. Kolb, Fakultät Biology, University of Konstanz, D-7750 Konstanz, Germany

The spectral intensity $S_J(f)$ of noise current generated by the transport of the hydrophobic ion dipicrylamine through lipid bilayer membranes was investigated at steady state. For a theoretical description of $S_J(f)$ the formalism derived for discrete transport mechanisms was used [1]. The transport of hydrophobic ions was approximated by a reaction scheme consisting of four constant ionic reservoirs (bulk phase and unstirred layer at the membrane-solution interface) and two fluctuating binding sites in the membrane near the interface. The voltage dependence of the rate constants for the adsorption to the membrane solution interface (k_{sm}), the translocation over the central dielectric barrier within the membrane (k_i) and the desorption in the aqueous solution (k_{ms}) was described due to the Eyring theory. The contribution of the ion diffusion between the bulk phase and the unstirred layer to $S_J(f)$ could be estimated. The model takes into account the non-zero membrane current obtained at nonequilibrium steady state after the time dependent diffusion polarization faded away. The measured shape of $S_J(f)$ is similar to that obtained under equilibrium conditions [2]. Between a frequency independent level of $S_J(f)$ at lower and higher frequencies the spectral intensity increases in proportion to the square of frequency. This slope of $S_J(f)$ is independent on voltage. The corresponding corner frequency is correlated with the translocation rate constant k_i. The measured voltage dependence of k_i could not be described by the Eyring theory. The high frequency level of $S_J(f)$ corresponds to the shot noise intensity of this transport system which decreases with increasing voltage. The decrease of $S_J(\infty)$ is correlated with the relation $k_{ms} \ll k_{sm}$. Within the used model the low frequency level of $S_J(f)$ is related to the time constants k_{ms} and k_{sm}. The following limits could be estimated for a diphytanoyl-lecithin/n-decane membrane in the presence of $3 \cdot 10^{-8} M$ dipicrylamin at equilibrium: $k_{ms} \leq 0.2 \ s^{-1}$ and $k_{sm} \geq 10^5 \ s^{-1}$. Both values increase with increasing voltage.

[1] E.Frehland (1978). Biophys. Chem. 8: 255-265
[2] H.-A. Kolb, P. Läuger (1977). J. Membrane Biol. 37: 321-345

NOISE ANALYSIS OF CARRIER-MEDIATED ION TRANSPORT UNDER NONEQUILIBRIUM CONDITIONS

H.-A. Kolb and E. Frehland, Fakultät Biology and Physics, University of Konstanz, D-7750 Konstanz, Germany

The noise current of carrier-mediated ion transport on bilayer membranes was measured for tetranactin Rb^+ under voltage clamp conditions. The shape of the spectral intensity $S_J(f)$ agrees under equilibrium and nonequilibrium conditions at steady state with the theoretically predicted behavior. The theory shows that the shot noise intensity of this ion transport mechanism yields to a frequency independent level of $S_J(f)$ at higher frequencies. The shot noise intensity decreases with increasing voltage. Both experimentally and theoretically it is shown that at nonequilibrium the Nyquist-theorem can no longer be applied for a derivation of the spectral intensity. Especially, the frequency independent tail of $S_J(f)$ at low frequencies is not proportional to the mean membrane conductance as was observed at equilibrium [1]. The increase of $S_J(f)$ between the low and high frequency limit occurs in a frequency range which is related to the corresponding relaxation time constants of the transport system. But in contrary to voltage jump current relaxation experiments where two relaxation times are derived the theory predicts for the noise analysis a third relaxation time constant of a non-zero contribution to $S_J(f)$ which is related to the interfacial reaction step. Besides the measurement of $S_J(f)$ the corresponding autocorrelation function $C_J(\tau)$ was determined. $C_J(\tau)$ shows a negative correlation for $\tau > 0$. Comparison of both theoretically equivalent methods of noise analysis shows that for this ion transport mechanism the determination of $C_J(\tau)$ is the less appropriate experimental approach due to principal methodical difficulties.

[1] H.-A. Kolb, P. Läuger (1978) J. Membrane Biol. 41, 167-187

1/f NOISE IN TRACK-ETCHED MICA MEMBRANES

H.-A. Kolb[1] and D. Woermann[2]

[1]Fakultät für Biologie, Universität Konstanz, 7750 Konstanz
[2]Institut für Physikalische Chemie der Universität Köln, 5000 Köln

The occurrance of the "1/f noise" in the spectral intensity generated by the passage of ions across biological and artificial membranes is a well known phenomenon but still unexplained[1]. Track-etched mica membranes yield under voltage clamp conditions to spectral intensities of 1/f shape. Application of these membranes offers a quantitative description of the 1/f noise current at different external parameters.

Clear muscovite mica membranes with a pore radius of about 100 nm, a pore density of about $10^6/cm^2$ and of variing thickness (10-50 μm) were prepared by the track-etch technique[2]. The membranes were placed between two aqueous KCl solutions at different concentrations ($c : 10^{-5}$M -1M KCl). The noise current could be described by the relation $S_J(f)$ $= A \cdot \overline{J}_m^2 \cdot f^{-\alpha}$, where $\overline{J}_m$ denotes the mean macroscopic current. The slope α was independent on electrolyte concentration, membrane thickness and temperature ($\alpha : 0.9-1.2$). An about constant value of α could be measured within a single experiment over a frequency range up to four orders of magnitude starting at 10 m Hz. For $c > 10$ mM A increases about proportional to the membrane resistance. For $c < 5$ mM A tends to increase with decreasing electrolyte concentration. In the range of $1mM \leq c \leq 10mM$ A is independent on voltage (V : 25-450 mV) whereas for $c > 50$ mM A shows an increase with increasing voltage. This increase of A is more pronounced for membranes at small thickness (10 μm).

[1]   B. Neumcke (1978) Biophys. Struct. Mechanism 4, 179
[2]   B. Klump, D. Woermann (1977) Ber. Bunsenges. phys. Chem. 81, 92

CURRENT FLUCTUATIONS IN BIOLOGICAL TRANSPORT SYSTEMS FAR FROM EQUILIBRIUM

Eckart Frehland, Fakultät für Physik, Universität Konstanz,
D-7750 Konstanz, Germany

A new theoretical approach to transport fluctuations around stable steady states in discrete biological transport systems [1],[2],[3] is used for the investigation of general fluctuation properties at nonequilibrium. An expression for the complex frequency dependent admittance at nonequilibrium is derived by calculation of the linear current response of the transport systems to small disturbances in the applied external voltage. It is shown that the Nyquist or fluctuation dissipation theorem, by which at equilibrium the macroscopic admittance or linear response can be expressed in terms of fluctuation properties of the system, breaks down at nonequilibrium. The spectral density of current fluctuations is decomposed into one term containing the macroscopic admittance and a second term which is bilinear in current. This second term is generated by microscopic disturbances, which cannot be excited by external macroscopic perturbations. At special examples it is demonstrated that this second term is decisive for the occurrence of excess noise e.g. the $1/f^2$-Lorentzian noise generated by the opening and closing of nerve channels in biological membranes. Furthermore the quadratic dependence on current (e.g. $1/f$ noise) comes out to be typical of excess noise.
We hope that this contribution will lead to a better general understanding of fluctuation properties as e.g. the shape of fluctuation spectra or the problem of minimization of noise in nonequilibrium processes. Both problems are discussed.

[1] E. Frehland, Biophys. Chem. 8, 255-265 (1978)
[2] E. Frehland, Biophys. Struct. Mechanism 5, 91-106 (1979)
[3] E. Frehland, W. Stephan, Biochim. Biophys. Acta 553, 326-341 (1979)

ION TRANSPORT THROUGH PORES: OSCILLATORY PHENOMENA AT TRANSIENT STATES

W. Stephan and E. Frehland, Fakultät für Physik, Universität Konstanz,
D-7750 Konstanz

This contribution is a theoretical analysis of time dependent ion transport through porous membranes. The principal aim is to give a foundation
for the interpretation of current relaxation experiments. Especially,
the influence of a feedback mechanism on the transport is considered. -
Two different models have been examined and compared: - 1. a discrete
jump diffusion model - The pore is looked upon as a sequence of n binding sites, separated by energy barriers. Due to electrostatic reasons
the following feedback mechanism is assumed [1] : Every pore can be occupied by no more than one ion ("one-ion model"). Mathematically the system is described by a linear n-dimensional system of differential equa-
tions with constant coefficients for the n independent occupation numbers N_i. It is shown that this system can be seen as a closed loop with
n+1 nodes - 2. a continuous diffusion model - The limit n → ∞ is carried out for regular energy profiles of a pore. A one-dimensional partial differential equation (Nernst-Planck equation) is derived from model 1. The feedback mechanism is taken into account by a special structure of the boundary conditions.
Results: If the jumps of the ions into or out of the pore are controlled
by the ionic concentrations at the pore mouths only, the transport will
be described by a spectrum of n relaxation times. If, however, there is
a linear connection between the number of pores already occupied and the
probability of ion entrance (linear feedback mechanism), the transport
system may exhibit damped oscillations. A necessary condition for that
is that the system will approach a stationary non-equilibrium state after the initial disturbance. For the oscillatory behaviour is valid: The
more the ion flux is directed and the more binding sites there are within
a pore, the smaller the damping will become. It can be shown that in the
limit n → ∞ (continuous model) even undamped oscillations might occur under extreme conditions.- Though this contribution is concerned with a very
special model, typical effects of ion transport through porous membranes
are described as consequence of interactions between the transport system
and the particles to be transported. A generalization of this model is
the well-known "Single file diffusion" model [2]. The analysis of this
more complex system, however, shows similar results [3].

[1] Frehland, E., Läuger, P., J. Theor. Biol. <u>47</u>, 189 (1974)
[2] Heckmann, K., Biomembranes <u>3</u>, 127 (1972)
[3] Frehland, E., Stephan, W. Biochim. Biophys. Acta <u>553</u>, 326 (1979)

WHY SINGLE FILE PORES ?

H.-H. Kohler
Lehrstuhl für Physikalische Chemie I
Universität Regensburg
D 84 Regensburg
Germany

There is strong experimental evidence that the potassium channels of nerve and muscle (1) and the gramicidin pores of artificial bilayers (2) are single-file pores (by "single-file pores" here, more precisely, "multi-ion single-file pores" are meant). The question arises, why nature might have chosen this type of permeation mechanism. Viewed from functional requirements, the potassium transport system of the nerve, for instance, is characterized by i) high permselectivity, ii) high transport rate, and iii) pronounced rectification properties. Comparing, under these aspects, the analytical results for saturated single-file pores with those obtained for a one-site pore, one finds specific advantages of the single-file concept which can be summarized as follows (cf. (1)):

Permselectivity: The permselectivity of the single-file pore depends on the energies of the barriers and on the energies of the sites. Thus, permselectivity can be controlled by binding selectivity.

Transport rate: Due to electrostatic repulsion between ions in a single-file pore, the maximum transport rate may be enhanced by several orders of magnitude.

Rectification: For a blocking particle to change site (such a particle here is assumed to be capable of entering into the pore but not of passing through), the pore content as a whole must shift. Therefore, the voltage dependence of the blocking (gating?) process is governed by the total charge of the pore content. For higher numbers of pore sites this leads to very steep rectification curves.

(1) Hille, B. & Schwarz, W. (1978). J. Gen. Physiol. 72, 409.
(2) Schagina, L.V., Grinfeldt, A.E. & Lev, A.A. (1978). Nature, Lond. 273, 243.

PROPERTIES OF IONIC CHANNELS MADE FROM DERIVATIVES OF GRAMICIDIN A

E. Bamberg and K. Janko, Fakultät für Biologie, Universität Konstanz,
D-7750 Konstanz

In the view of the single stranded model for the Gramicidin A channel
it was stated in the literature that the desformylated form of Gramicidin A is inactive as a channel former. A carefully cleaned desformyl Gramicidin A, however, shows a strong activity. The mean life time
of the channel and its amplitude is remarkably reduced compared to
unmodified Gramicidin A. In a series of experiments the ionic specificity of the desformyl Gramicidin channel was measured. Especially
the blocking phenomena of the thallous ion was studied and compared
with the effect on normal Gramicidin A.
By N-O-acyl rearrangement a desethanolamine Gramicidin was synthesized and its voltage dependence for channel formation was studied.
Surprisingly the molecule shows at low pH, when the free carboxylic
group on the C terminal is protonated, no voltage dependence, whereas at high pH, when the carboxylic group is deprotonated, a normal
voltage dependence occurs. This phenomenon is discussed on the basis
of the molecular model of the Gramicidin channel.

STATE DEPENDENT BLOCKING EFFECTS ON THE ALAMETHICIN PORE BY DIVALENT
CATIONS

W. Hanke and G. Boheim
Lehrstuhl Zellphysiologie, Ruhr-Universität Bochum, Postfach 102148,
D-4630 Bochum 1

Free-diffusion theories have been found inadequate for describing a variety of properties of ionic fluxes in biological channels. For this
reason deviations from independent movement of ions have been discussed
in terms of competitive binding of different ions to discrete sites
within a channel. In case of the gramicidin channel, where the inner
wall is simply lined by carbonyl oxygens, related phenomena have been
described. By way of example, the permeability to alkali cations is
blocked in dependence on the concentration of divalent cations to which
the gramicidin channel is impermeable.

A different type of ionophore, alamethicin, which induces pores in lipid bilayer membranes, too, adopts several distinct conductance states.
The values of state conductances vary over up to three orders of magnitude. In case of 3 M KCl (21 °C) the lowest conductance state is found
at 40 - 45 pS depending on the lipid used. However, with 2 M $CaCl_2$ this
state could not be resolved (< 4 pS) which seems to indicate that it is
virtually impermeable to Ca^{++} (and Cl^-). Experiments with mixtures of
KCl and $CaCl_2$ show that the permeability of the lowest conductance state
to K^+ can be blocked to a great extent by Ca^{++}, whereas the highest pore
state conductances approach proportionality to bulk solution conductivity.

The present picture of alamethicin pore structure is that of a pore of
variable diameter. Values for the different pore state diameters have
been estimated which range from ∼ 6 Å to > 15 Å. Thus the alamethicin
pore gives the possibility to estimate the critical pore diameter (in
case of simple pore structures) above which independent ion movement
through a pore occurs and below which this approximation seems to be
incorrect.

STRUCTURAL REQUIREMENTS FOR MEMBRANE MODIFYING ACTIVITY IN ALAMETHICIN-
TYPE ANTIBIOTICS

H. Brückner, G. Jung
Institut für organische Chemie, Universität Tübingen, Auf der Morgenstel-
le, D-7400 Tübingen

W. Hanke, G. Boheim
Lehrstuhl Zellphysiologie, Ruhr-Universität Bochum, Postfach 102148,
D-4630 Bochum 1

In order to elucidate the relations between structure and membrane modifying properties of alamethicin-type antibiotics investigations with
natural, modified natural and synthetic analogues have been carried out.
The activity pattern splits up into three modes: 1. activity in form of
resolvable, stable ($\tau \succ$ 1 ms) pores, 2. activity in form of voltage-dependent burst formation (not resolvable fluctuations) which finally leads
to membrane lysis, 3. inactive behaviour. Whereas modes 2. and 3. are to
differentiate by hemolysis and artificial lipid bilayer measurements,
single pore events (mode 1) are only observable with the bilayer system.

Conformational analysis of the pore former (mode 1) alamethicin established (a) a helical part of 10 - 12 residues starting at the N-terminus
with (b) a hydrophilic amino acid (GLN) in between, (c) a flexible part
around GLY which is found near the middle of the molecule, followed in
a short distance by (d) a helix breaking PRO and (e) β-turns in the C-
terminal part. The lacking of GLN in the α-helical part (-b, a conven-
tionally synthesized nonadecapeptide analogue) or of GLY in the middle
position (-c, the natural analogue trichotoxin A40 highly purified by
Craig distribution) leads to active compounds showing not resolvable
fluctuations (mode 2). The lacking of both (-b -c, a synthetic hexade-
capeptide analogue) results in complete loss of activity (mode 3). Selective cleavage of the AIB-PRO bond yielded an inactive N-terminal do-
decapeptide (-d -e, α-helical) and an inactive C-terminal hexapeptide
(-a -b -c, with β-bends). The natural analogue suzukacillin, on the
other hand, which meets all the above mentioned structural requirements,
exhibits stable pore formation (mode 1).

INVESTIGATIONS ON MONOLAYERS OF ALAMETHICIN-TYPE ANTIBIOTICS AT AN AIR/WATER INTERPHASE

S. Überschär and G. Boheim
Lehrstuhl Zellphysiologie, Ruhr-Universität Bochum, Postfach 102148,
D-4630 Bochum 1, F.R.G.

Different alamethicin-type antibiotics were spread at an air/water interface and pressure/area (π/A)-curves recorded using a Langmuir surface balance (Lauda trough) and the Wilhelmy plate technique (Fromherz through). For at least 10 - 15 min a stable monolayer is formed without appreciable loss of molecules into the subphase (KCl solutions of different ionic strength, pH 5.8 - 6.0). In another series of experiments the antibiotics were dissolved in the subphase and its surface adsorption was pursued. In a very slow diffusion controlled process up to approximately 1/3 to 1/2 of the dissolved molecules were transferred into the surface after ~ 24 h. This is consistent with circular dichroism observations that the alamethicin-type antibiotics are found in two different conformations: 1. with a small α-helix content of 10 - 20 % and relatively good water solubility and 2. with a large α-helix content of 40 - 50 % and relatively poor water solubility.
Alamethicin and suzukacillin exhibit a smooth $\mathbf{\L}$-shaped $\tilde{\pi}$ /A-characteristic which approaches saturation without collaps at 35 - 37 dyn cm^{-1} with 1 mM KCl. The corresponding area per molecule is calculated to be ~ 2.5 nm^2. Trichotoxin A40, however, shows a phase transitions at 9 dyn cm^{-1} and saturates at 30 - 32 dyn cm^{-1}. A larger phase transition range is found with antiamöbin I and samarosporin at 3 dyn cm^{-1}, whereby the π /A-characteristic saturates at 24 - 26 dyn cm^{-1}. In lipid bilayer membranes alamethicin and suzukacillin induce conductance fluctuations by pore formation, whereas with pure Trichotoxin A40 the induced current fluctuations could not be resolved. Antiamöbin I and samarosporin are found to be inactive.

TRANSPORT OF MATTER THROUGH MEMBRANES. THE INFLUENCE OF PROTEIN ON THE KINETICS OF THE FREE DIALYSIS OF DETERGENTS

H. Craubner

Max Planck-Institut für Züchtungsforschung, Cologne

The influence of protein (Bovine serum albumin, BSA) on the kinetics of the free dialysis (against zero detergent conc., Visking membranes) of detergents (Sodium dodecyl sulfate, SDS; Lubrol; Sucrose palmitate-stearate) was investigated with regard to the time course and the concentration dependence. The SDS-BSA system was studied in details at 22 $^{\circ}$C. In the case of pure detergent-sodium phosphate buffer solutions (0.01 m; pH 7; 1.5 x 10^{-3} m EDTA) the permeation experiments resulted in simple exponential decrease of the SDS-concentration, approaching zero within less than 24 hours for initial SDS-concs.< 1 % by wt. If protein was present additionally, a more complex exponential superposition behaviour of the permeation time course was observed. Here, zero detergent concentration was not   reached within more than 72 hours. In this context, for constant zero or finite protein concentration, the permeation or dialysis coefficient (λ), i.e. the reciprocal permeation time constant (τ), decreased linearly with increasing SDS concentration. With regard to the dependence on the protein concentration, however, an exponential decrease of λ was oberved.

The theoretical permeation and diffusion analysis was carried out on the basis of nonequilibrium thermodynamics. Here , above the critical micelle concentration (cmc ≈ 0.1 wt.-% SDS) , the dynamical multiple equilibria with regard to the detergent micelles and the protein-detergent micelles were taken into consideration besides the amphiphile dissolved in monomeric form. By taking account of further measurements, as e.g. density, partial specific volumes, viscosity, detergent binding (at 0.25 wt.-% SDS: 1 mol. BSA binds approx. 100 mols. SDS) etc., the influences of the various Donnan effects were studied.

The results obtained are of practical interest e.g. with regard to detergent removal and reconstitution after amphiphile solubilization of biomembranes and membrane proteins[1].

./.

---

1   H. Craubner, F. Koenig, and G. H. Schmid, Z. Naturforsch. 30c,615 (1975); 32c, 384(1977).

VESICLE SPREAD MEMBRANES, A NOVEL WAY FOR MEMBRANE RECONSTITUTION

Hansgeorg Schindler, Biozentrum, University of Basel, CH-4056 Basel, Switzerland

Lipid or lipid-protein vesicles disintegrate at the air-water interface and form a monolayer[1]. Such monolayers were combined to true bimolecular lipid-protein membranes[2]. The monolayer selfassembly has been analyzed in detail[3]. Here we present the criteria for successful membrane formation. Examined parameters were: a. monolayer surface tension (measured during membrane formation), b. size, c. concentration of vesicles, d. type of lipid, e. lipid/protein ratio, f. preciseness and pretreatment of the membrane frame. There are three basic requirements for

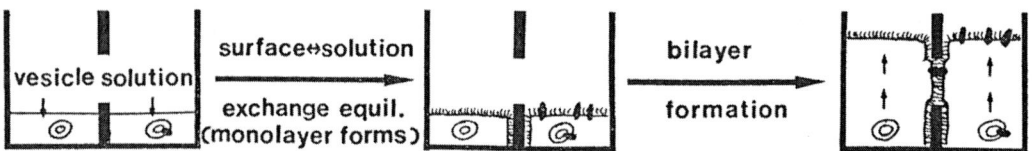

membrane formation from homogeneous vesicle solutions (cf. Fig.): 1) The equilibrium surface pressure p_e must exceed a value p_e^c of 20 dynes/cm (membrane frame pretreated with hexadecane or vaseline). Hydrocarbon solvent free membranes (frame treated with triglycerides) can be formed at surface pressures above 35 dynes/cm. Such p_e values (up to 40 dynes/cm) were reached by increasing the vesicle size in agreement with our theoretical analysis[3]. 2) When the water levels are raised during membrane formation (see Fig.) the monolayer area increases and p decreases. Above a vesicle concentration of 0.5 mg/ml the exchange between vesicles and monolayer is fast enough[3] to maintain p above p_e^c. 3) Perfectly planar hole boundary with irregularities less than 1 μm (diameter .1 to .4 mm, frame thickness less than 12 μm); for manufacturing cf. ref. 4.

1. R. Verger and F. Pattus (1976) Chem. Phys. Lipids 16, 285-291.
2. H. Schindler and J.P. Rosenbusch (1978) Proc. Natl. Acad. Sci. U.S. 75, 3751-3755.
3. H. Schindler, Biochim. Biophys. Acta, in press.
4. H. Schindler and G. Feher (1976) Biophys. J. 16, 1109-1113.

MATRIX PROTEIN/ATPase PROTEOLIPID/ACETYLCHOLINE RECEPTOR IN VESICLE SPREAD MEMBRANES

Hansgeorg Schindler, Biozentrum, University of Basel, CH-4056 Basel, Switzerland

Matrix Protein (MX) from E. coli (with J. P. Rosenbusch, Biozentrum Basel).
A.) Vesicle spread membranes could be formed directly from outer membranes of
E. coli. The observed single channel data are identical to those published for isolated and vesicle reconstituted MX^1, whereas channel inactivation is faster and
occurs at lower membrane potentials.   B.) MX triplets were incorporated into vesicles at molar ratios of up to 10^9 lipids/MX triplet. The number of triplet channels
observed is close to the number expected if every protein is active. Irreversible
aggregation to larger channel arrays via lateral diffusion (10^{-9} cm^2/sec estimated)
can be followed.

H^+ATPase Proteolipid (P) from yeast mitochondria (with N. Nelson, Biozentrum
Basel). Butanol extracted P was incorporated into lipid vesicles. Vesicle spread
membranes show proton conductance (selectivity ratio $H^+/K^+ > 800$). The current-
voltage relation is slightly nonlinear. At pH 2 (ether lipids were used) single
channels of about 10 pS H^+ conductance are observed. Experiments on the molecul-
arity of the channel are in preparation.

Acetylcholine Receptor (R) from Torpedo (with U. Quast, Biozentrum Basel).
Closed vesicles obtained from Torpedo electric organ were spread into a monolayer.
Surface pressure is about 20 dynes/cm and the lipid/R ratio is of the same order as
in the vesicles. R vesicles were mixed with pure lipid vesicles for dilution and
surface pressure increase. Vesicle spread planar bilayers (R density typically
10^{10} R/cm^3) show capacitance and resistance values close to those of pure lipid
membranes. Agonists induce conductance which is curare sensitive in agreement
with the established pharmacology of R. No agonist induced conductance was observed using pretoxinated vesicles. Single channels show several milliseconds lifetime and about 20 pS conductance. The maximum number of channels observed
approximates that estimated from the monolayer studies taking desensitization into
account.

1. H. Schindler and J. P. Rosenbusch (1978) Proc. Natl. Acad. Sci. U.S. 75,
   3751 - 3755.

FORMATION AND CHARACTERIZATION OF BILAYER MEMBRANES FROM VESICLE-SPREAD MONOLAYERS

A. Schneider and E. Bamberg, Universität Konstanz, Fachbereich Biologie, D-7750 Konstanz

Lipid bilayers were formed from monolayers by spreading lipid vesicles on an air-water surface. The formation of the bilayers can be improved by applying an osmotic shock to the vesicles.

The monolayers were formed in the two chambers of a teflon cell and combined to a Montal-Mueller-membrane by the usual technique. The hole in the teflon foil (130 μm diameter) used for membrane formation was pretreated with Vaseline. No organic solvent was present. The method has been described in a similar way by Schindler and Rosenbusch (2). The specific capacitance of the bilayers was measured using a voltage-step method. The results for some lipids are given in Table 1. Fig. 1 shows the relation between lecithin chain length and thickness of the hydrocarbon interior of the membrane. The membranes were also characterized by means of Gramicidin A single channel measurements. Former investigations of Hladky and Haydon on the relation between mean channel duration and membrane thickness were used to estimate the thickness of Diphytanoyl-PC membranes at 25°C. The result was in good agreement with the membrane thickness calculated from the capacitance measurements (see Fig. 1).

The technique, which is reported here, could be used for reconstitution of membrane proteins into lipid bilayer membranes.

Table 1. Specific capacitace values for vesicle-membranes

Lipid	number of measurements	spec.cap.cm² μ Farad
PC 24:1	6	0.414 ± 0.036
PC 22:1	7	0.697 ± 0.073
PC 18:1	7	0.798 ± 0.044
PC 16:0	15	0.559 ± 0.082
Asolectin	11	0.530 ± 0.035
egg-PC	6	0.693 ± 0.038
Phosphatidyl-serin	7	0.649 ± 0.047
Diphytanoyl-PC	8	0.840 ± 0.031

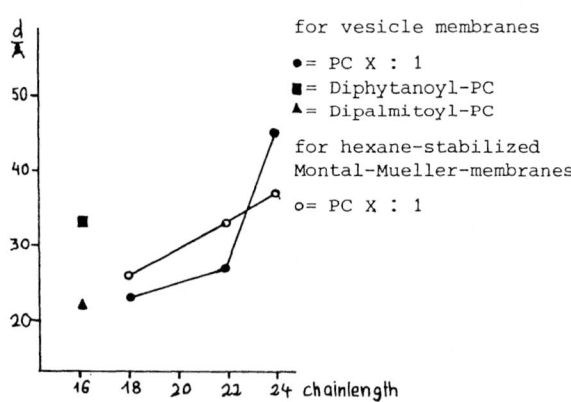

for vesicle membranes
● = PC X : 1
■ = Diphytanoyl-PC
▲ = Dipalmitoyl-PC

for hexane-stabilized Montal-Mueller-membranes
o = PC X : 1

Fig. 1. Thickness of membrane hydrocarbon interior versus lecithin chainlength. Membrane thickness was calculated with $\mathcal{E}$ = 2.1

1) R. Verger and F. Pattus; (1976) Chem. Phys. Lipids 16, 285

2) H. Schindler and J.P. Rosenbusch; (1978) P.N.A.S. 75, 3751

STUDIES ON PHOTOCURRENT KINETICS OF PURPLE MEMBRANE ON BLACK LIPID
MEMBRANES

A. Fahr, E. Bamberg, Fachbereich Biologie, Universität Konstanz
D-7750 Konstanz

Purple membrane fragments were added to one aqueous phase of a positively charged black membrane. After approximately 30 min. the membrane
became photoelectrically active. Recent investigations (1,2) showed,
that purple membrane fragments are attached to the membrane in an oriented manner.
Transient current of this system, caused by 572 nm dye laser flashes,
are measured and sampled. A typical time course of this photocurrent
is shown in Fig. 1

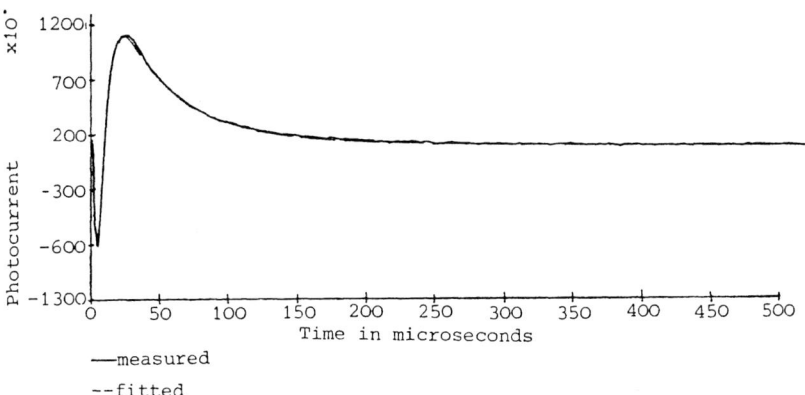

Fig. 1: Time course of photocurrent (aqueous phase:
pH = 6.9, 0.1 M NaCl, Tris-Maleat 5 mM
fitted: $\tau 1$ = 4.2 microsec., $\tau 2$ = 42.5 microsec.,
$\tau 3$ = 370 microsec., τ of measuring system = 0.6 µsec.)

This curve was fitted with a kinetic equation consisting of three ex-
ponential terms: $I(t) = - a_1 \cdot \exp(t/\tau_1) + a_2 \cdot \exp(t/\tau_2) + a_3 \cdot \exp(t/\tau_3)$.
It was possible to correlate the time constant τ_2 with the M-product
of the bacteriorhodopsin-photocycle. This is confirmed by determination
of the transient photocurrent time constants at different temperatures.
It came out that the activation energy for the M-product time constant
is the same as obtained by flash photometric experiments on purple membrane in solution (3).

1) Karvaly, B. and Dancshazy, FEBS Letters 76, 36-40 (1977)
2) Bamberg et al., Biophys. Struct. Mechanism, in press
3) Dencher and Wilms, Biophys. Struct. Mechanism 1, 259-271 (1975)

ASSOCIATION EQUILIBRIA BETWEEN OLIGOMERS OF BAND 3-PROTEIN FROM HUMAN
ERYTHROCYTE MEMBRANES: TIME NEEDED FOR RELAXATION TOWARDS EQUILIBRIUM

H.-J. Dorst, G. Pappert, D. Schubert, and R. Benz[+], MPI für Biophysik,
D-6000 Frankfurt a.M. and [+]Fachbereich Biologie, Universität Konstanz,
D-7750 Konstanz, FRG

Band 3, the main integral protein of the human erythrocyte membrane, was solubilized
and purified in high concentrations of acetic acid, and afterwards the organic solvent was exchanged against aqueous solutions by dialysis. It was demonstrated by
analytical ultracentrifugation that, under appropriate conditions, the solubilized
protein exists in a monomer/dimer/tetramer association equilibrium. We have suggested
that band 3 will show this type of association also in the intact erythrocyte membrane[1].

Analysis of the shape of the boundaries in sedimentation velocity experiments on solubilized band 3 demonstrates that the time needed for relaxation towards association
equilibrium is different for different band 3-samples and varies between < 1 min
("rapid") and > ~h ("slow"). The type of relaxation shown by a sample was found to
be dependent on the donor of the red cells used for membrane preparation and may thus
be of genetic origin.

To exclude that the effects observed may be artifacts due to irreverisble denaturation
of the protein during isolation, band 3 was also prepared, instead by the use of acetic
acid, by selective extraction of erythrocyte membranes by the nonionic detergent Am-
monyx-LO[2]. The results obtained with these samples were very similar to those described above, both with respect to the occurrence of an association equilibrium and to the
dependency of boundary shape and thus of relaxation time on the blood donor.

Interaction of band 3-samples which, in aqueous solution, showed "slow" relaxation,
with black lipid membranes (BLM) led to an increase in BLM conductivity proportional
to the power (3.0 ± 1.0) of protein concentration. This is consistent with the previous suggestion that the conducting sites are induced by band 3-tetramers[3]. It further shows that, in the BLM, an association equilibrium exists which is rapid or intermediate on the time scale of the BLM experiment. Thus, the relaxation times char-
acterizing the association equilibrium of band 3 seem to be shifted to lower values
in a lipid bilayer as compared to aqueous solutions.

1. Dorst, H.J. and Schubert, D. (1978) Abstract H4, Annual Meeting of the German
    Biophysical Society, Ulm; (1979) submitted for publication.

2. Yu, J., Fischman, D.A. and Steck, T.L. (1973) J. Supramol. Struct. 1, 233-248.

3. Bleuel, H., Wiedner, G. and Schubert, D. (1977) Z. Naturforsch. 32 c, 375-378.

PROPERTIES OF LARGE ION-PERMEABLE PORES FORMED BY PORINS FROM SALMO-
NELLA TYPHIMURIUM IN LIPID BILAYER MEMBRANES

R. Benz and T. Nakae, Fachbereich Biologie, Universität Konstanz,
D-7750 Konstanz and Tokai University, School of Medicine, Bohseidai
Isehara 259-11, Japan

Porin trimers from Salmonella typhimurium were isolated by detergent
solubilization and chromatography. The addition of small protein concentrations (10^{-8}M to 10^{-12}M) to the aqueous phases bathing artificial
lipid bilayer membranes causes a strong increase of the membrane conductance up to 4-5 orders of magnitude. The membrane conductance is a
linear function of the protein concentration. The conductance increase
in the presence of porin is strongly dependent on the type of the lipid
used for membrane formation. In the case of lipids with a pronounced
polar head group (phosphatidylserine, phosphatidylcholine and phosphatidylethanolamine) the influence on the conductance in the presence of
the different porin trimers (34K, 35K and 36K molecular weight of the
monomers) is relatively low (two orders of magnitude), whereas it is
high for membranes from monoolein and oxidized cholesterol. When complex I (oligomers with a molecular weight of 1 million Dalton) was
used instead of the trimers the conductance was three to ten times
lower.

The addition of small amounts of protein to membranes of small area
results in a stepwise increase of the conductance. The single increment of this increase is independent on the type of the lipid used for
membrane formation and has a value of 2.5 nS for 1 MKCl. Under the
assumption of a pore length of 7 nm, the diameter of the pore is estimated to be about 1.5 nm. This value which has to be considered as a
lower limit is consistent with earlier permeability studies [1]. The
data presented here are consistent with the hypothesis that one porin
trimer forms one pore [2] and not three pores [3].

[1]   T. Nakae, Biochem. Biophys. Res. Commun. 64, 1224-1230 (1975)
[2]   J.M. DiRienzo, K. Nakamura, and M. Inouye, Ann. Rev. Biochem. 47,
      481-532 (1978)
[3]   H. Nikaido, Angew. Chem. 91, 394-407 (1979)

REVERSIBLE ELECTRICAL BREAKDOWN OF LIPID BILAYER MEMBRANES

R. Benz and U. Zimmermann, Fachbereich Biologie, Universität Konstanz, D-7750 Konstanz and Institut für Biophysikalische Chemie, Kernforschungsanlage Jülich, D-5170 Jülich

Charge pulse experiments were performed with lipid bilayer membranes from oxidized cholesterol/n-decane at relatively high voltages (several hundred mV). The membranes show an irreversible mechanical rupture if the membrane is charged to voltages in the order of 300 mV. In the case of the mechanical rupture the voltage across the membrane needs about 50 µs-200 µs to decay completely to zero. At much higher voltages, applied to the membrane by charge pulses of about 500 ns duration, a decrease of the specific resistance of the membranes by nine orders of magnitude is observed (from $10^8 \Omega cm^2$ to $0.1 \Omega cm^2$) without a mechanical rupture of the lipid bilayer membrane. Due to the high conductance increase (breakdown) of the bilayer it is not possible to charge the membrane to a larger value than the critical potential difference V_c. For 1 M alkali ion chlorides V_c was about 1 V. The temperature dependence of the electrical breakdown voltage V_c is comparable to that being observed with cell membranes [1]. V_c decreases between $2^\circ C$ and $48^\circ C$ from 1.5 V to 0.6 V in the presence of 1 M KCl.

In order to study the influence of the charging time on the absolute value of V_c charge pulse experiments were performed on artificial lipid bilayer membranes in the time range between 10 ns and 10 µs. Between 300 ns and 5 µs at $25^\circ C$ and between 100 ns and 2 µs at $40^\circ C$ V_c showed a strong dependence on the charging time of the membrane and decreases from 1.2 V to 0.5 V ($25^\circ C$) and from 1 V to 0.4 V ($40^\circ C$). For other charging times below and above these ranges the breakdown voltage seems to be constant. The results indicate that the breakdown phenomenon occurs in less than 10 ns.

The pulse length dependence of the breakdown voltage is consistent with the interpretation of the electrical breakdown mechanism in terms of the electro-mechanical model. However, it seems possible that below a charging time of the membranes below 300 ns ($25^\circ C$) and 100 ns ($40^\circ C$) other processes (like the Born energy) are initiated.

[1]   Zimmermann, U., Pilwat, G., Beckers, F., Riemann, F.
      Biolectrochem. Bioenerg. 3, 58 (1976)

DEPENDENCE OF THE ELECTRICAL BREAKDOWN VOLTAGE ON THE CHARGING TIME IN
VALONIA UTRICULARIS

U. Zimmermann and R. Benz, Institut für Biophysikalische Chemie, Kernforschungsanlage Jülich, D-5170 Jülich and Fachbereich Biologie, Universität Konstanz, D-7750 Konstanz

Charge-pulse experiments were performed on giant algal cells of Valonia
utricularis. For a charging time of 420 μs the breakdown voltage is
about 750 mV (18°C), a value that is in close agreement with earlier
results obtained with current pulses [1]. If the membrane is charged to
the breakdown voltage in a shorter time, the breakdown voltage is found
to be a function of the duration of the charge pulses. Whereas towards
smaller pulse lengths down to 10 μs only a small, but significant, increase in the breakdown voltage is observed (1.1 V at 10 μs pulse length
and 18°C), a strong increase in the breakdown voltage is found for even
shorter charging times. For a pulse length of 800 ns the breakdown voltage has a value of about 2.4 V (18°C) and a plateau seems to be reached
for a pulse duration of 500 ns. For charge pulses of 1 to 2 μs duration
the breakdown voltage decreases from 3.6 V at 3°C to 1.6 V at 25°C by
more than a factor of two.
Voltage relaxation studies in the low-field range suggest that the time
constants of the two membranes arranged in series, tonoplast and plasmalemma, are similar. From this, it is suggested that both membranes
show electrical breakdown, whereby the breakdown voltage of a single is
probably half the value of the total breakdown voltage. Its dependence
on pulse length is therefore considered to be an intrinsic property of
one single membrane. The strong dependence of the breakdown voltage on
the charging time of the membrane further supports the interpretation
of the breakdown phenomenon on the basis of the electro-mechanical model proposed earlier. In this model it is assumed that the electrical
and mechanical compressive forces are counterbalanced by elastic re-
storing forces within the membrane. However, towards very short pulses
(less than 800 ns), where a plateau seems to be reached, other processes
may be generated by the application of the electric field. We discuss
whether one of these processes is the ion movement through the membrane
induced by a high electric field (Born energy).

[1]   Coster, H.G.L., Zimmermann, U.
      J. Membrane Biol. 22, 73 (1975)

INTERACTION OF BURN TOXIN WITH LIVER CELLS AND ARTIFICIAL LIPID BI-LAYER MEMBRANES

K. Schmidt and R. Benz, Chirurgische Universitätsklinik, 7400 Tübingen and Fachbereich Biologie, Universität Konstanz, 7750 Konstanz

Isolated murine skin was burnt in a standard high temperature burn model ($250^{\circ}C$, 20 sec). From the burnt skin a lipid protein complex was isolated which exhibited toxic activities after intraperitoneal injection into acceptor mice [1]. Metabolic and ultrastructural studies using rat liver perfusion and isolated hepatocytes as target systems revealed alterations of cellular and especially of mitochondrial membranes and a reduced adaptability in gluconeogenesis and urea synthesis after stimulation with amino acids and lactate. After lipid extraction by organic solvents intravenous injection of the apo-protein into rabbits caused platelet aggregation and an increase of activity of lysosomal enzymes in serum.

Lipid bilayer experiments were performed in the presence of the lipid containing burn protein as well as of the lipid free apo-protein. In the first case almost no influence of the protein (concentration ca. 10^{-6} g/ml) on the electrical properties (conductance ca. 10^{-8} Scm^{-2} in 0.1 M NaCl) of the artificial lipid bilayer membranes was found, and no increase of the effect by the addition of ionic detergents (SDS and cholate 10^{-3} %) could be observed. In the presence of the lipid free protein (concentration 10^{-6} g/ml) conductance between 10^{-5}-10^{-6} Scm^{-2} were measured. The addition of 0.1 % SDS to the protein stock solution increased the effect of the lipid free protein considerably and conductance up to $2 \cdot 10^{-3}$ Scm^{-2} could be observed (0.1 M NaCl, 10^{-6} mg/ml protein, 10^{-3} % SDS), whereas 10^{-3} % SDS alone showed no influence. The conductance increase of lipid bilayer membranes in the presence of lipid free burn protein from different preparations showed a strong correlation to the biological toxicity, being higher for samples of a higher toxicity.

[1] Schölmerich, J., Kremer, B. and Schmidt, K. (1977) Acta Chir. Aust. 9, 159-166

DEVELOPMENT OF A NEW COULTER COUNTER SYSTEM:

MEASUREMENT OF THE VOLUME, INTERNAL CONDUCTIVITY, AND DIELECTRIC BREAKDOWN

VOLTAGE OF A SINGLE GUARD CELL PROTOPLAST OF VICIA FABA AND TUMOR CELLS

U. Zimmermann, G. Pilwat, H. Schnabl[*]

Institute of Chemistry (Biophysical Chemistry)
Nuclear Research Center Jülich
P.O.Box 1913, 5170 Jülich, W.-Germany

[*]Institute of Botany and Microbiology
Technical University Munich
Arcisstr. 21, 8000 Munich, W.-Germany

Abstract

A new Coulter Counter is described which measures on single microscopic cells the volume, membrane breakdown voltage and the underestimation of volume after breakdown, this last parameter reflecting a part the internal conductivity of the cells. The system requires very few cells for measurement and does not require the population to be normally distributed in volume or free of debris from the preparation of the cells.

The new development is to apply a voltage ramp across a Coulter orifice through which a particle flows. A differential dual orifice system is used to reduce the magnitude of the amplified ramp signal to that of the signal produced by the particle in the presence of a DC voltage across the orifice. Breakdown is detected by an underestimation of the cell volume once a critical voltage across the cell membrane has been established.

Measurements made on protoplasts isolated from guard cells of Vicia faba with a mean volume of 1800 μm^3 reveal a breakdown voltage of 1 V and an average volume underestimation of about 30% after breakdown. Variations in this underestimation reflect the different internal structure of the protoplasts in terms of the size and number of non-conducting compartments. Different stages of vacuole development exist in each preparation. The breakdown voltage of 1 V suggests that only membrane breaks down and that the electro-mechanical properties of these protoplast membranes are similar to other cell systems, when the pulse length dependence of the breakdown voltage is taken into consideration.

Breakdown measurements on tumor cells BT3C strain established from fetal brain cells of rats show a similar breakdown behaviour as observed with protoplasts. It seems to be possible to distinguish between various phases of the growth cycle due to different breakdown voltages of the membrane and of different internal conductivities.

GUARD AND MESOPHYLL CELL PROTOPLASTS OF VICIA FABA:PLASMALEMMA ULTRASTRUCTURE

J. Vienken, H. Schnabl[*], U. Zimmermann, H. Ziegler[*]

Institute of Chemistry (Biophysical Chemistry)
Nuclear Research Centre Jülich
P.O.Box 1913, 5170 Jülich, W.-Germany
[*]Institute of Botany and Microbiology
Technical University Munich
Arcisstr. 21, 8000 Munich, W.-Germany

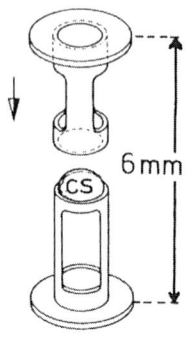

6 mm

Fig. 1: Modified
Leybold specimen
holder

We investigated protoplasts of guard and mesophyll cells of Vicia faba by means of the freeze etching technique. The cell suspension (cs) was frozen in melting N_2 using a modified Leybold Bioetch 2005 specimen holder (Fig. 1).

Freeze-fractures of the plasmalemma of guard and mesophyll cell protoplasts of V. faba demonstrate that the inner core of the plasmalemma exposed by the freeze fracturing process exhibits areas with distinct, highly organized structures. Between areas of intramembraneous particles dispersed randomly on a relatively smooth fracture face, membrane domains showing an extremely regular planar hexagonal array of particles are interspersed (Figs. 2 and 3). The dimensions of the whole hexagonal lattice are about 0.5 µm in diameter, the center-to-center spacing of the lattice-unit is about 22 nm and the particle size about 9 nm. The particles in the hexagonal arrays are accompanied by complementary pits in the opposite fracture face of the plasmalemma.

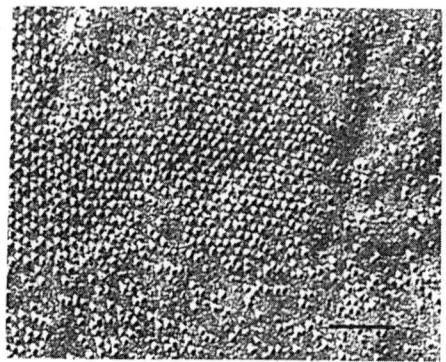

Fig. 2: Regularly arranged particles in the plasmalemma of mesophyll protoplasts of V. faba.
Bar: 0.1 µm.

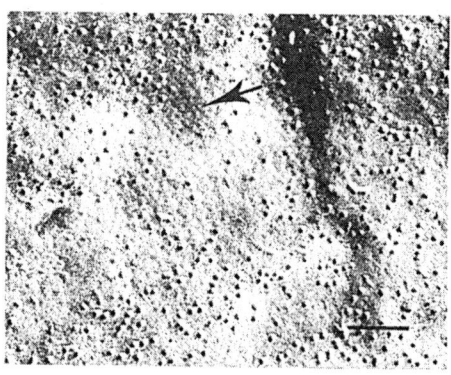

Fig. 3: Hexagonal arranged pits (arrow) in the opposite fracture face of the plasmalemma of mesophyll cells of V. faba.
Bar: 0.1 µm.

X-RAY DIFFRACTION AND ELECTRONMICROSCOPIC STUDIES

OF THYLACOID MEMBRANES

N. Hodapp, W. Welte, D. Walter, W. Kreutz

Institut für Biophysik und Strahlenbiologie
der Universität Freiburg im Breisgau
Albertstrasse 23, D-7800 Freiburg, W. Germany

Wet pellets of the photosynthetic apparatus obtained from thylacoid preparations of Rhodopseudomonas viridis were exposed to X-rays with the membranes specifically oriented. The orientation of membrane stacks made it possible to distinguish between stack scattering and isotrope background scattering.

The least square fit procedure of Welte (+) was used to evaluate the electrondensity profile. This procedure fits the scattering intensity of model stacks applying the Shannon-scanned points of the experimental intensity. The model stack is described by Fourier series and Gauss-shaped distance statistics. Two modifications of the Welte procedure were necessary to improve both the quality and the uniqueness of the fit.

1) The statistical distribution of the number of double membranes is described by:

$$N(n) = e^{-\frac{n}{n_0}}$$

were N is the number of stacks with n double membranes per stack and n_0 the mean value.

2) Because $n_0 \approx 4$ in our case the difference between the mean electrondensity of the stacks and the mean electrondensity of the sample has to be taken into account. This was done by introducing the 0-th coefficient into the Fourier series description of the membrane profile.

The evaluation of the experimental X-ray diffraction pattern by the modified procedure led to a unique determination of the thylacoid membrane electrondensity profile. When the above information is considered along with the watercontent of the sample, an approximative determination of the contribution of transmembrane protein to the electrondensity profile is possible.

A comparison of the statistical behaviour of membrane distances obtained from both X-ray diffraction and freeze-etch samples showed that it is not possible to use electronmicroscopically determined statistical data for the evaluation of X-ray patterns.

(+) W. Welte, W. Kreutz, Adv. Polym. Science 30 (1979) 161

ON THE TRANSMEMBRANE ELECTRICAL POTENTIAL DIFFERENCE IN CHLOROPLASTS

STUDIED BY ELECTROCHROMISM

R. Tiemann, D. DiFiore and H.T. Witt
Max-Volmer-Institut für Physikalische Chemie und Molekularbiologie
Technische Universität Berlin
Straße des 17. Juni 135, 1000 Berlin 12

The importance of the transmembrane electrical potential difference for the energy requirement of the phosphorylation is not yet clarified, because controversial results concerning the amount and nature of this potential are given in the literature. Therefore an attempt was made to solve this controversy by two differnt approaches:

1) The electrochromic absorption change at 520 nm was studied under steady state illumination as a function of the concentration of monovalent ions (KCl). Varying the KCl concentration in the range from 0.5 to 100 mM a monotone decrease of the steady state transmembrane potential from 80 to nearly zero mV was found. The absolute calibration was achieved by comparison with the amplitude of electrochromism in single turnover flashes which generate a transmembrane voltage of 50 mV. In the cited range of concentrations no salt dependence of the amplitude was found in single turnover flashes, whereas the decay time was affected. The simplest theoretical approaches to describe the salt dependence of membrane potentials are given by Donnan and Gouy-Chapman. In both theories the charge density (space or surface) is an essential parameter, which is changed in chloroplasts by proton binding. Thus the steady state proton uptake was measured via the absorption change of the pH-indicator cresol red ($7 < pH < 9$) or bromcresol purple ($5 < pH < 7$). This uptake seems to be independent on the concentration of KCl. Comparison of the measur-ed with theoretical data lead to a fairly good fit by a potential of the Donnan type.

2) The membrane conductivity of the charged ionophore complexes nonactin K^+ and valinomycin $\cdot K^+$ was measurded as a function of pH ($pH_{out}=pH_{in}$), preillumination ($pH_{out} \neq pH_{in}$, pH_{in} variable), both as a function of the concentration of monovalent ions (KCl, KCl + NaCl). In the case where $KCl=10$ mM, $\sigma_1=0.0005$ $e^-/Å^2$, $\sigma_2=0.0015$ $e^+/Å^2$ the Gouy-Chapman theory predicts the ratio of the conductivities $G_1/G_2=66$. The Donnan theory says in the comparable case $G_1/G_2=3.5$. The measurements yield under these conditions: $G(pH=9)/G(pH=6)=3.5$ and $G(I_h^{min})/G(I_h^{max})=3.6$, i.e. the Donnan theory seems to be more suitable to describe the present data.

The consequences of the results will be discussed.

STUDY OF THE ELECTRON- AND PROTON TRANSPORT
IN INSIDE-OUT THYLAKOIDS

R. Tiemann, G. Renger and P. Gräber
Max-Volmer-Institut für Physikalische Chemie und Molekularbiologie
Technische Universität Berlin
Straße des 17. Juni 135, 1000 Berlin 12

Inside-out thylakoids are an attractive system for the investigation of the functional sidedness of the thylakoid membrane, especially for the study of protolytic reactions coupled with wateroxidation which occur at the inner side of normal thylakoids. The functional arrangement of the electron transport chain of these particles is not yet clarified. Therefore, comparative pulse spectroscopic investigations of the following components of the chain in normal and inside-out thylakoids have been performed: P 680, X 320, PQ (Plastoquinone) and P 700. It was found that the inside-out thylakoids are slightly enriched in PS II (P680/P700 $\gtrsim$ 1) and that the connection between the PQ-pool and PS I (P 700) is interrupted. Accordingly, these particles mainly behave as PS II - vesicles. The effect of trypsin on inside-out thylakoids was studied in a similar way as was done before on normal chloroplasts (1). In contrast to normal chloroplasts the reoxidation kinetics of X 320$^-$ in inside-out vesicles is not affected by trypsination. This result provides a further proof that the membrane is turned inside-out (former proofs 2,3,4). Surprisingly, the watersplitting enzyme system Y seems to be rather stable against trypsination which is shown by oxygen and proton measurements. In inside-out particles the overall half rise time of the proton release due to wateroxidation is found to be 1 ms which is in accordance with measurements on normal chloroplasts. Preliminary results on the oscillatory pattern of this proton release will be discussed.

1) Renger, G. and Tiemann, R. (1979) Biochim. Biophys. Acta 545, 316-324

2) Anderson, B., Akerlund, H.E. and Albertsson, P.A. (1977) FEBS Lett. 77, 141-145

3) Tiemann, R. et al. Biophysiktagung Ulm (1978) Poster F1

4) Gräber, P., Zickler, A. and Akerlund, H.E. (1978) FEBS Lett. 96, 233-237

INITIAL KINETICS OF ATP-SYNTHESIS AND OF CONFORMATIONAL CHANGES IN THE CHLOROPLAST ATPase STUDIED BY EXTERNAL ELECTRIC FIELD PULSES

E. Schlodder and H.T. Witt

Max-Volmer-Institut für Physikalische Chemie und Molekularbiologie
Technische Universität Berlin
Straße des 17. Juni 135, 1000 Berlin 12

Applying external electric field pulses to a Chloroplast suspension an electric potential difference is generated across the energy - transducing membrane. By this artificial membrane potential ATP - synthesis (1) and the energy-dependent exchange of adenine nucleotides tightly bound to the ATPase indicating a conformational change of the ATPase can be induced (2).

The rise and decay time of the transmembrane potential difference measured by the field indicating absorption changes (3) is in the order of 10 μs. This time reflects the charging and discharging of the membrane capacitance. With respect to the mechanism of the energy transduction it is of interest to measure the dependence of ATP form- tion and the exchange of tightly bound nucleotides on the time of energzation of the membrane. In contrast to excitation by light pulses the excitation by external electric field pulses offers the opport- unity to adjust the time of energization in the range of 100 μs up to 60 ms and membrane potentials up to 200 mV.

The following results were obtained: 1) A linear rate of ATP formation has been observed without any initial lag. The rate depends on the transmembrane potential difference. 2) Only a fraction of the [14]C - nucleotides tightly bound to the ATPase is exchanged on energization, i.e. only a fraction of ATPases is activated and changes their con- formation. The exchange can be described by first-order kinetics with a half-life time of 2.0 - 2.5 ms independent on the membrane potential. The size of the fraction depends on the magnitude of the transmembrane potential difference. 3) The initial rate of the exchange is practically identical with the rate of ATP synthesis.

These results support the proposed model (2) that the rate of ATP synthesis is controlled by the fraction of active ATPases and not by the variation of the turnover time.

(1) H.T. Witt, E. Schlodder and P. Gräber (1976) FEBS Lett. 69,272
(2) P. Gräber, E. Schlodder and H.T. Witt (1977) BBA 461,426
(3) E. Schlodder and H.T. Witt (1978) Biophysik-Tagung Ulm, Poster F2

RAPID KINETICS OF ATP-SYNTHESIS IN CHLOROPLASTS BY ACID/BASE TRANSITION

G.H.Schatz, E.Schlodder, M.Rögner and P.Gräber
Max-Volmer-Institut für physikalische Chemie und Molekularbiologie
Technische Universität Berlin
Straße des 17. Juni 135, 1ooo Berlin 12

ATP-synthesis by the membrane-bound ATPase of chloroplasts is coupled
to a transmembrane protonflux. This protonflux can be induced artifi-
cially by acid/base transitions (1,2). Kinetic studies of the resulting
ATP-synthesis require conditions, where the energization, ΔpH, is not
altered by the protonflux.
For this purpose we have set up ΔpH-pulses of defined energization
(2 $\leqslant$ pH $\leqslant$ 5) and of defined duration (5 ms $\leqslant$ t $\leqslant$ 3oo ms) by rapid acid/
base mixing techniques. Under these controlled conditions the depend-
ence of the time-resolved ATP-synthesis on both, the extent of energization, ΔpH, and - with constant energization - on the absolute pH-
values has been studied. We present the following experimental results:

- ATP-synthesis starts with a time-lag of about 5 ms.
- ATP-formation requires a ΔpH of at least 2,1 and simultaneously
  a minimal internal proton-concentration of about 10^{-6} M.
- the dependence of the rate of ATP-formation on ΔpH is sigmoidal.
  Saturation is reached at the same maximal rate as obtained by
  other methods of membrane energization.
- ATP-synthesis is influenced by the absolute values of the external pH and is decreased if $pH_{ext} \geqslant 8,3$.

The interpretation of these results is based on a model, which proposes
a constant turnover time and a variable active fraction of the ATPases
to control enzyme activity (3) .
It is proposed that the time-lag is identical with the turnover time.
Indeed similar values (4-5 ms) have been determined in independent
experiments (4,5). The influence of the transmembrane energy gradient
on the ATP-synthesis is discussed under kinetic and thermodynamic
aspects.

References: (1) A.T. Jagendorf & E.Uribe
                Proc.Natl.Acad.Sci., USA 55,17o (1966)
            (2) B.Huchzermeyer & H.Strotmann
                Z.Naturforsch. 32c, 8o3 (1977)
            (3) P.Gräber,E.Schlodder & H.T.Witt
                Biochim.Biophys.Acta 461, 426 (1977)
            (4) E.Schlodder &H.T.Witt
                Biophysik-Tagung Konstanz (1979)
            (5) G.H.Schatz, E.Schlodder & P.Gräber
                Biophysik-Tagung Ulm, Poster F 12 (1978)

DEMONSTRATION OF ELECTRICAL SURFACE CHARGE EFFECTS AT THE INNER SIDE OF THE THYLAKOID MEMBRANE

Hans-Ludwig Huber and Bernd Rumberg

Max-Volmer-Institut, Technische Universität, D-1000 Berlin 12, FRG

Thylakoid vesicles which are formed by the thylakoid membrane are the place of the primary events of photosynthesis in green plants. Electrical charges on the surface of membranes give rise to surface potentials which depend on both surface charge density and salt concentration in the surrounding solution. In this work we present data from functional measurements which demonstrate the influence of surface charges at the inner side of the thylakoid membrane.

Photosynthetic electron transport is sensitized by two special photoactive chlorophyll-a molecules which are series connected. The electron transfer from $chl-a_{II}$ to $chl-a_I$ is coupled to the release of protons into the inner phase of the thylakoids. Therefore the recovery time of photooxidized $chl-a_I$ is controlled by the proton concentration in the inner phase and can serve as an intrinsic indicator of the proton concentration at the inner membrane-solution interface (decrease of the inner pH from 8 to 5 results in a 10-fold slow-down of the recovery time).

The recovery time of photooxidized $chl-a_I$ is also dependent on the salt concentration in the medium (decrease of monovalent salt concentration from 300 to 3 mM at an outer pH of 8 results in a nearly 10-fold slow-down). If we assume a surface charge of negative sign at the inner side of the thylakoid membrane we can attribute this effect of low salt to a decrease of the pH at the inner membrane-solution interface due to an increase of the negative surface potential. The experimental results in the range of outer pH from 4.5 to 8 and salt concentration from 2 to 300 mM can all be interpreted in this way quantitatively on the basis of the Gouy-Chapman surface potential theory. The consequences can be summarized as follows:

1. The surface charge density at the inner side of the thylakoid membrane is around $2 \cdot 10^{-3}$ $e^-/\text{Å}^2$ at outer pH values between 6 and 8. The corresponding surface potential amounts to -118, -63 and -25 mV at 3, 30 and 300 mM monovalent salt concentration resp.
2. The isoelectric point at the inner surface is around pH 4.5.
3. The above given data are compatible with the assumption that exclusively lipids are responsible for the surface charge at the inner side of the thylakoid membrane.

Ammonium Transport in the plasma membrane of Riccia fluitans

Hubert Felle

Abteilung Biophysik der Pflanzen

Institut für Biologie I

D-7400 Tübingen 1

The plasma membrane of Riccia fluitans may be described by means of an equivalent circuit with a proton pump as reasonable current source in parallel with the sum of the other ion channels in question. There is little doubt, that potassium is the dominating ion.

In the concentration range of 1-20 μmol, ammonium depolarizes the Riccia membrane within a few seconds from -230 mV to -110 mV, which is close to the diffusion potential. Repolarization occurs immediately after exchange with ammonium free medium.

At maximal depolarization, the conductivity of the membrane is increased by roughly 60%, whereas only 15% were measured, testing the hexose-transport system, which is similar in electrical phenomena. For ammonium, K_m-values of 1-2 μmol were calculated, when plotted $1/\Delta E_m$ vs $1/NH_4$. The same experiments with methylamine however yielded values one order of magnitude higher, which is in good agreement with flux measurements using ^{14}C-methylamine as marker for NH_4.

The uptake of ammonium can easily be split into a carrier component for low concentrations, and a diffusional one for higher NH_4-concentrations. This is underlined by comparison of transport rates with and without CN^- at NH_4-concentrations higher than 50 μmol.

Flux data and electrical data are highly potassium- and potantial dependent, to a lesser extend to external pH. Presenting I-V-curves, a possible NH_4-transport system in close connection to the potassium channel is discussed.

EFFECT OF ANTI-MICROTUBULE HERBICIDES ON Ca^{2+} TRANSPORT AND ENERGY TRANSDUCTION IN PLANT MITOCHONDRIA

Cornelia Hertel and Dieter Marmé
Institut für Biologie III
University of Freiburg
D-78 Freiburg (W.-Germany)

Agents like ORYZALIN, TRIFLURALIN and AMIPROPHOSMETHYL are known to cause aberrant deposition of cellulose microfibrils controlled by micro-tubules in Oocystis solitaria and to induce flagellar shortening in Chlamydomonas reinhardii. It has been suggested, that changes of the intracellular Ca^{2+}level are involved. Mitochondria are known to accumulate large amounts of Ca^{2+} which can be released upon addition of suitable effectors.

We have investigated the effect of various anti-microtubule herbicides on Ca^{2+} accumulation into mitochondria isolated from corn and squash. All the herbicides tested, show an inhibition of Ca^{2+}uptake into mito-chondria (Fig.1). The molecular mechanisms of the herbicide action has been investigated (Fig.2). Trifluralin and Oryzalin interfere with the energy supplying system, Amiprophosmethyl seems to influence directly the transportsystem. These results support the assumption that anti-microtubule herbicides achieve their effect by increasing cytoplasmic Ca^{2+} concentration.

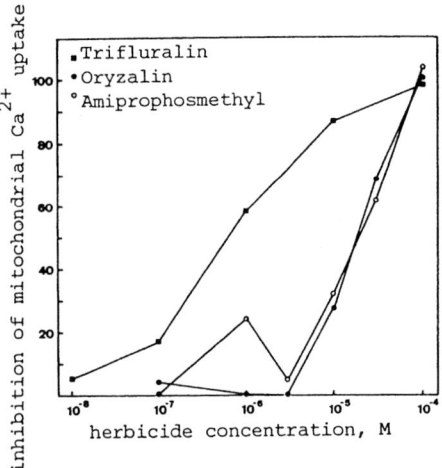

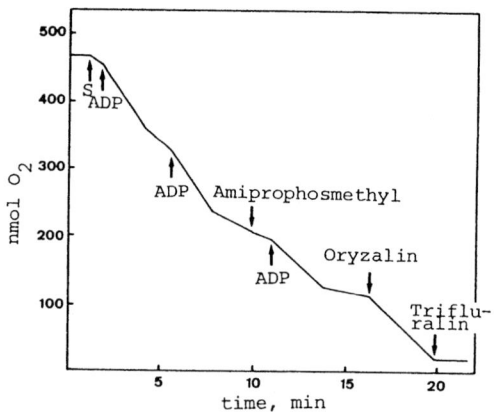

Fig.1: Inhibition of Ca^{2+}uptake by different herbicide-concentrations

Fig.2: Influence of herbicides on mitochondrial O$_2$-consumption.

EVALUATION OF THE ELECTRONDENSITY PROFILE OF THE FROG
ROS-DISK MEMBRANE IN VIVO USING X-RAY DIFFRACTION

J. Funk, I. Wutschel, W. Welte, W. Kreutz

Institut für Biopyhsik und Strahlenbiologie
der Universität Freiburg im Breisgau
Albertstrasse 23, D-7800 Freiburg, W. Germany

The structure of the ROS-disk membrane was studied on living narcotized frogs. A mathematical unique solution for the electrondensity profile could be determined. The unique determination of the solution was made possible by two factors:

1) A generalized theory for the evaluation of X-ray diffraction patterns of Welte (1) regarding

   a) statistical distribution of the membrane distances,

   b) electrondensities in the inter- and intradisk spaces which differ from the mean electron-density.

2) An exact measurement of the isotrope background scattering; this became practicable since the ROS-cells are strictly oriented towards the center of the pupil.

Proof of living state of the frogs was given by the fact that half an hour after an experiment all frogs showed quite normal reflexes.

In previous papers (2) it was often assumed that all rhodopsin molecules are concentrated on the outer face of the disc membrane. Because of the asymmetry of the derived electrondensity profile this interpretation becomes rather improbable.

distance distributions

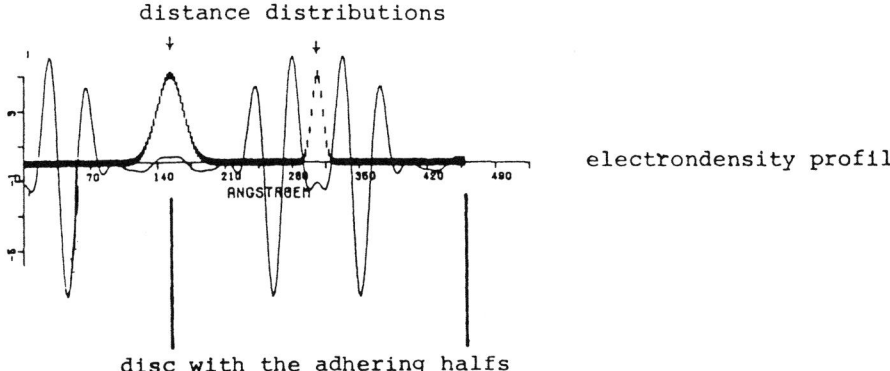

electrondensity profil

disc with the adhering halfs
of the interdiscspace

1) Welte, W., Kreutz, W., Adv. in Polymer Science (1979) 30, 161

2) Corless et al., Exp. Eye Res. (1976) 23, 295-324

SURFACE POTENTIALS IN INTACT CATTLE ROD OUTER SEGMENTS

P.P.M.Schnetkamp[$], U.B.Kaupp[$] and W.Junge[$]

Max-Volmer-Institut (PC14), Technische Universität Berlin
Strasse des 17.Juni 135, 1000 Berlin 12, Germany

Bovine rod outer segments are isolated in such a way as that they behave as a two-compartment system. Stacked disks are embedded in the cytoplasm, which is separated from the external medium by an intact outer envelope (1). Disk membranes contain a large intracellular binding capacity for calcium ions (as well as other divalent cations), whic is predominantly located inside disks and at which divalent cations can be reversibly replaced by protons (2). Photoexcitation of rhodopsin is followed by a release of some fraction of bound calcium ions, probably to the intradiskal space (3). In these experimen free calcium concentrations in all compartments are intercommunicated by the addition of the ionophore A23187. It was concluded that these observations are not easily reconciled with the calcium-transmitter hypothesis of Yoshikami and Hagins (4). This communication is dedicated to the interrelation between the binding capacity for divalent cations in disk membranes and the surface potential at disk membranes.

A first approach uses the photolytic properties of rhodopsin as an intrinsic pH-indica tor. In the presence of A23187 the local pH at the membrane surface appears to be decreas when all divalent cations are complexed by EDTA. Subsequent addition of manganese ions results in an increase of the local pH by 1.0-1.1 pH-unit. Without A23187 only marginal effects are observed when the divalent cation concentration in the external medium is changed. Intrinsically permeant cations (ammonium ions added as acetate salt) also increa the local pH, but without the necessity of adding an ionophore. The here reported changes of the local pH at the membrane surface are not affected by the addition of protonophores and are not transient effects.

A second approach is based on the distribution of the amphiphilic indicator dye neutra red (positively charged acid form predominates in the membrane phase) between the membrane phase and the external medium. Within the concentration range of neutral red used the amount of neutral red present in the membrane phase increases linearly with the external concentration of neutral red. The partition coefficient increases by a factor of 11-17 (dependent on the bleaching state of the membrane) when manganese ions are removed from the membrane by EDTA in the presence of A23187.

We interpret these observations by the decrease of the surface potential, due to the presence of fixed negative surface charges, by cations, and by the influence of the surface potential on the local pH at the membrane surface and on the partition of a positively charged amphiphilic probe respectively. Both approaches (which necessitate different assumptions) lead to an estimated change of the surface potential at rod disk membranes of 60-70 mV upon replacement of protons by manganese ions at the binding sites of the dis membranes.

Binding of divalent cations to rod disk membranes appears to affect the surface potential. Conversely, the surface potential appears to affect the rapid light-induced release of bound calcium ions rather strongly. Permeant cations (ammonium ions added as acetate salt) abolish rapid calcium release at a concentration of 20 mM. When the light-regulated binding sites are directly exposed to the medium by fragmentation of intact rod outer segments, rapid calcium release is abolished by 10 μM streptomycin (a threefold positively charged aminosugar). Finally, rapid calcium release is abolished when disk membranes are dissolve in detergents (triton X-100, nonylglucose). These observations suggest that the binding sites of rod disk membranes are located at the membrane surface.

[$]present address:

Lehrstuhl für Biophysik
Universität Osnabrück
Postfach 4469
D-4500 Osnabrück
Germany

References:

(1) Schnetkamp, P.P.M., u.a., BBA552(1979)37
(2) Schnetkamp, P.P.M., BBA, in press
(3) Kaupp, U.B., u.a., BBA 552 (1979) 390
(4) Yoshikami, S. & Hagins, W.A., Biophys. J
    10 (1971) 60a

ON THE PROTEIN COMPOSITION OF BOVINE ROD OUTER SEGMENT DISK MEMBRANES

Rainer Uhl
MPI für biophysikalische Chemie, Am Fassberg, D-3400 Göttingen, FRG
Nancy Semple, Jack Pasternak
Biology Dept. University of Waterloo, Ontario, Canada
Tom Borys & E.W. Abrahamson
Chemistry Dept. University of Guelph, N1G 2W1, Ontario, Canada

The pigment molecule rhodopsin is generally believed to account for 90% of the total disk membrane protein[1]. This figure has recently been questioned by Siebert et al.[2], who report the existence of three major protein peaks on SDS-PAGE scans, all in the molecular weight region of 30 000 to 42 000. One of the proteins can be washed out, the remaining two, however, are intrinsic. If this peak separation were not artefactual and if only one of the peaks were rhodopsin, it certainly could not constitute 90% of the total disk membrane protein. We have therefore attempted to confirm Siebert's data and to unequivocally identify rhodopsin among the disk protein components.

Purified disk membrane preparations (A_{280}/A_{500} ca. 1.85), when subjected to SDS gel electrophoreses according to the method of Lämmli, usually yield two clearly separated major protein bands of 34 500 and 37 500 MW respectively. Like Siebert et al. we find the staining intensity ratio of the two peaks H (higher MW) and L (lower MW) to vary considerably, but only when the proteins are delipidated before electrophoreses. Without delipidation the ratio is fairly constant, ca. 2:1 in favour of H. To identify rhodopsin on the gels we have used three rhodopsin properties:
1) rhodopsin contains the chromophore retinal covalently bound
2) rhodopsin is a glycoprotein
3) rhodopsin is phosphorylated upon illumination.
We have therefore taken ROS membranes, incubated them with ^{32}P-ATP in the light, reduced the chromophore with the mild, not denaturing reagent NaCNBH$_3$, and then performed SDS gel electrophoreses. The gels were first scanned at 330nm to identify the reduced chromophore, then stained with PAS to identify the glycoprotein, and subsequently superstained with coomassie blue. The clearly distinct coomassie blue peaks were then sliced out of the gel and counted for radioactivity. It was found that by all three criteria only one of the peaks, namely H, was rhodopsin. This would indicate the existence of a second major protein fraction in the disk membrane other than rhodopsin.

Ref. 1   Heitzmann, H. (1972) Nature New Biol. 235 , 114
     2   Siebert, F., Schmid, H., Mull, R.H. (1977) BBRC 75 , 1071

STRUCTURAL CONSEQUENCES OF MAGNESIUM ATP-ase ACTIVITY IN BOVINE ROD OUTER SEGMENT DISK MEMBRANES

Rainer Uhl
MPI für biophysikalische Chemie, Am Fassberg, D-3400 Göttingen, FRG
Tom Borys and E.W. Abrahamson
Dept. of Chemistry, University of Guelph, N1G 2W1, Ontario, Canada

Fresh bovine rod outer segment (ROS) preparations exhibit relatively high Mg-ATPase activity (initial rate: 400 nmole ATP hydrolysed min^{-1} mg rhodopsin^{-1}). The enzyme activity is accompanied by a dramatic change in light-scattering from the ROS: in 5 min the total light-scattering decreases by about 50%. The angular dependence of this dark light-scattering increment suggests that the underlying event is a change in refractive index.

Incubation of ROS with Mg^{2+} and ATP not only causes the light-scattering to decrase in the dark, but also is prerequisite for yet another light-scattering change, namely a rapid (msec range) flash-induced further decrease in turbidity. This light-response we have labelled "A_L", as opposed to the dark effect of Mg^{2+} and ATP: "A_D". The connection between the three phenomena: ATPase activity, "A_D" and "A_L" appears to be established by the following findings:

Conditions, which inhibit Mg-ATPase activity, (100μM DCCD, 100% inhibition; 100μM oligomycin, 50% inhibition; AMP-PNP, competitive inhibition depending on the ratio ATP/AMP-PNP) also inhibit both "A_D" and "A_L" to the same extent. Other conditions, which leave ATPase activity unaffected, but disallow "A_D", consequently block "A_L". When applied after "A_D" has already taken place, they reverse it and also block "A_L". These conditions include the Cl^-/OH^- carrier tributyltin, the K^+ ionophore valinomycin in the presence of K^+ and the zwitterionic anion PIPES, and low concentrations (0.001%) of the detergent Triton X-100. A third class of inhibitors, namely mM amounts of Ca^{2+} and 100μM chlorpromazine, leave both ATPase activity and "A_D" unaffected, but suppress the light response.

On the basis of these observations we conclude that there is a Mg-ATPase system in the photoreceptor disk membrane, which "energises" the disk in the dark, thus enabling it to exhibit a rapid structural light-response. Both the enabling process and the light-response are structural phenomena, which are accompanied by marked light-scattering increments.

Ref. 1   Uhl, R., Borys, T., Abrahamson, E.W. (1978) Biophys. J. 21 13€
     2   Uhl, R., Borys, T., Abrahamson, E.W. (1979) Photochem Photobic
          29 703

FLASH SPECTROSCOPIC STUDIES ON ELECTROPHYSIOLOGICALLY INTACT BOVINE RETINAE

Spalink, J.-D.

Institut für Neurobiologie der Kernforschungsanlage Jülich GmbH

Postfach 1913

D-5170 Jülich

Although solubilized bovine rhodopsin and suspensions of bovine rod outer segments have often been studied by flash spectroscopy, up to now no such experiments were performed with intact bovine retinae.

To approach physiological conditions, almost completely dark adapted retinae were taken from male or female cattle, blindfolded for at least one hour before enucleation of the eye, which took place immediately after killing of the animal. At that time the blood pressure was still at the physiological value.

A definite part of the retina (area centralis) was prepared, carefully placed in a cell and superfused with physiological saline. This cell allows simultaneous recording of both the transretinal electroretinogram and the spectroscopic transmittance changes. At $37^{\circ}C$, the R_1 and R_2 component of the early receptor potential and the a-wave of the late receptor potential could be detected after an exciting flash of 500 nm wavelength, bleaching 8-10% of the rhodopsin.

The flash is delivered by a tuneable dye laser with a pulse width of about 6 ns, causing a flash artifact in the spectroscopic part of the registration unit of max 40 µsec. This is due to the bandwidth, which ranges from DC to 20 kHz (-3dB). These relatively short flash artifacts allow studies of the kinetics of the decay of lumirhodopsin and the rise of MI at temperatures from $1^{\circ}C$ to about $15^{\circ}C$ and also the decay of MI or the rise of MII in the range of $1^{\circ}C$ to $37^{\circ}C$ in the superfused retina.

The influence of light scattering has been compensated as good as possible by a proper choice of the space angle under which the cell is seen by the photomultiplier (EMI 9801 B).

CHEMICAL CONTROL OF MEMBRANE TRANSPORT BY THE ACETYLCHOLINE RECEPTOR SYSTEM

Julius Bernhardt and Eberhard Neumann

Max-Planck-Institut für Biochemie, D-8033 Martinsried/München, BRD

The function of the acetylcholine receptor system in excitable membranes of nerve and muscle cells is to control the passive transport of Na^+ and K^+ ions, which are common carriers of bioelectric signals. Details of this gating function can only be studied by the investigation of the transport properties of membrane-bound receptors. A powerful technique at a less complex level than whole cells is the measurement of metal ion exchange fluxes induced by activator molecules in sealed, receptor-rich membrane fragments (1). The aim of these experiments is to derive from flux parameters information on the permeability control system. Recently, the basic elements of a rigorous physical chemical analysis of complex flux curves have been discussed in terms of explicit relations to receptor-specific rate constants (2,3). In particular, the ligand-dependent reduction (inactivation) of the flux (analogous to pharmacological desensitization) has been analyzed in terms of a specific reaction scheme, excluding alternatives. The results are principally the same as the data from electrophysiological studies on muscle and electroplax: the activated, ion-conducting conformation of the acetylcholine receptor is a transient, short-lived metastable state; comparable to the conducting state of the Na^+ ion gating system in axonal excitable membranes (3). A variety of experimental correlations to the electrical properties of whole cells shows that molecular-functional details of receptor-mediated transport can be studied with sealed membrane fragments.

(1) Kasai, I. and Changeux, J.P. (1971), J. Membrane Biol. 6, 1-23.

(2) Bernhardt, J. and Neumann, E. (1978), Proc. Natl. Acad. Sci. USA, 75, 3756-3760.

(3) Neumann, E. (1979) in: Molecular Mechanisms of Biological Recognition (ed. Balaban, M., Elsevier North Holland, Amsterdam) pp. 245-265.

DISCRETE CURRENT FLUCTUATIONS PRODUCED BY SINGLE K+-CHANNELS
IN THE SQUID AXON MEMBRANE

E. Neher and F. Conti
Max-Planck-Institut für biophysikalische Chemie, Göttingen, BRD
and
Laboratorio de Cibernetica e Biofisica (CNR), Camogli, Italy

Ionic currents in the electrically excitable membrane of nerves are composed of contributions from many independent ion-channels. Fluctuation analysis has shown that each channel contributes approximately 1 pA of current, when in the open state (1,2). The method of patch clamp recording (3) allows currents through biological membranes to be measured with sufficient resolution to resolve unitary events of this magnitude. The method has so far been employed to record single channel currents of the acetylcholin receptor-channel in muscle membrane. In this report, a modification is described which allows high-resolution measurements to be made on the squid axon membrane:

A fire-polished L-shaped micropipette with 1-2 μm opening, introduced longitudinally into the axon interior, is pressed sideways against the internal membrane surface. Both the pipette and the axon interior contain low ionic strength perfusion medium. Electrical connections and electronic circuits are arranged, such that voltage clamp type membrane currents can be measured from the small patch of membrane, covered by the pipette opening. In order to obtain the best possible resolution we selected conditions in which the normal K^+-gradient across the membrane is inverted, i.e. high external K^+-concentration (460 mM), and low internal ionic strength with internal K^+ virtually absent. We also suppressed contributions from Na^+-channels by the addition of TTX(0.3 μM). Under these conditions the voltage sensitivity of channel gating is preserved, but is slowed down in its time course. The membrane "resting"-potential is +60 to +90 mV (inside positive).

The following sequence of events is observed in the patch recording when the membrane potential is gradually shifted from its "resting" value to more negative values: At or close to +60 mV the record shows only small fluctuations, mainly due to background noise sources ($\approx 0.1 pA_{rms}$). This is expected since at this potential channels are open (or inactivated), and the electrochemical driving force for K^+-ions across the membrane is small. As the membrane potential approaches the range in which voltage dependent gating occurs (+20 to -20 mV), large fluctuations develop ($\approx 10 pA_{rms}$) which again become small at -20 to -25 mV. In the latter voltage range the fluctuations consist of discrete, one-sided and inwardly directed waveforms. The large fluctuations around 0 mV represent fluctuations in the number of open channels which are in a voltage dependent, dynamic equilibrium between the open and closed state. At -20 to -25 mV this equilibrium is shifted so much to the closed side that only occasionally a fluctuation appears which, then, probably represents the current contribution from an individual ionic channel. These data can give valuable information on the underlying molecular events, although resolution is marginal at the present time.

(1) Conti, F., DeFelice, L.J., Wanke, E. 1975 J. Physiol. 248: 45-82
(2) Begenisich, T., Stevens, C.F. 1975 Biophys. J. 15: 843-846
(3) Neher, E., Sakmann, B. 1976 Nature 260: 799-802

THE EFFECT OF HIGH EXTRACELLULAR POTASSIUM ON THE KINETICS OF

POTASSIUM CONDUCTANCE OF THE SQUID AXON MEMBRANE

W. Stühmer and F. Conti

Fachbereich Physik E17, Technische Universität, München,BRD
and
Laboratorio di Cibernetica e Biofisica (CNR),Camogli,Italy

K currents were studied under voltage clamp in intact axons or in axons perfused intracellularly with solutions containing 300 meq-K^+, sucrose and phosphate ions at pH 7.3. The axons were superfused with extracellular solutions containing: (460 - x) mM-NaCl (or TMACl); x mM-KCl; 50 mM-$CaCl_2$; .5 μM-TTX; 5 mM-TrisCl buffer; pH 7.8; $0 \leq x \leq 460$. Changes in the extracellular K^+-concentration, $[K]_0$, changed the kinetics of K conductance, g_K, as measured by the slope (for outward currents) of the instantaneous I-V characteristic of K currents.

1. Increasing $[K]_0$ hastens the increase of g_K produced by prepulses of variable duration, t_p, from the standard holding potential of (-70 mV) to a membrane potential, E_p, of 40 mV to 100 mV. The asymptotic value, $g_K(\infty)$, reached by g_K at these depolarizations was independent of $[K]_0$, but the rise time of g_K (from 0 to $.9g_K(\infty)$) was roughly halved, for E_p=50 mV, in going from $[K]_0$=10 mM to $[K]_0$=460 mM.

2. Increasing $[K]_0$ slows down the decrease in g_K following membrane repolarization. This effect was demonstrated by measuring the time course of inward K currents produced by stepping the membrane potential to values, E_t, between -120 mV and -40 mV, after the prepulses. The half-time, $t_{\frac{1}{2}}$, of the decline of tail currents was found to increase with $[K]_0$ and/or with t_p. The dependence of $t_{\frac{1}{2}}$ on t_p could be mainly attributed to potassium accumulation in the outer Schwann space(1) and was observed also for very low $[K]_0$. However, for $[K]_0$=460 mM we observed a significant increase of $t_{\frac{1}{2}}$, upon increasing t_p from 1 ms to 5 ms, which could not be ascribed to potassium accumulation and was correlated with $g_K(t)$. For a fixed t_p, upon increasing $[K]_0$ from 10 mM to 460 mM, $t_{\frac{1}{2}}$ increased by a factor of 3 to 5 for E_t=-60 mV and by a factor of 1.5 to 2.5 for E_t=-120 mV.

3. The above effects were all reversible within the time (~5 minutes) required to complete a series of measurements in high $[K]_0$, and return to the control solution ($[K]_0$=0, or $[K]_0$=10 mM). However, prolonged and/or repetitive exposures to high $[K]_0$ produced also progressive and poorly reversible effects, such as a slowing down of the whole g_K kinetics and a stronger dependence of $t_{\frac{1}{2}}$ on t_p. Since axons were kept at their resting potentials between g measurements, these effects could also be the consequence of prolonged depolarizations.

The short term effects described above are all consistent with the idea that high $[K]_0$ decreases the rate constant of the transition from the open to closed state of K channels. It has been proposed that the presence of extracellular K^+ is a prerequisite for the occurrence of the closed to open transition in the node of Ranvier(2). Our data are better described by assuming that open K channels are multi-ion single-file pores(3) which cannot switch to the closed configuration from a state of high ion occupancy.

(1) Frankenhaeuser,B., Hodgkin,A.L. 1956 J. Physiol. 131: 341-376
(2) Dubois,J.M., Bergman,C. 1977 Pflügers Arch. 370: 185-194
(3) Hille,B., Schwarz,W. 1978 J. Physiol. 72: 409-442

EFFECT OF GABA ON NON-SYNAPTIC POTASSIUM CHANNELS IN CRAYFISH MUSCLE

W. Finger and J. Dudel

Physiologisches Institut der Technischen Universität München

Biedersteiner Strasse 29, 8000 München 40

The application of excitatory transmitter substances to postsynaptic membranes in vertebrate and invertebrate muscle induces the synaptic membrane current to fluctuate around its mean value due to the random opening and closing of ionic channels located in the postsynaptic membrane. The statistical analysis of this current noise yields information about the properties of an individual synaptic channel in terms of its conductance γ and its mean lifetime τ (cf. Neher and Stevens 1977). Recently we have started to analyse current noise produced by application of γ-aminobutyric acid (GABA) to inhibitory synapses on small crayfish muscle fibres (Dudel et al. 1977; Finger 1979). The synaptic inhibition plays an important role at information transmission processes in the central nervous system and has not been investigated so far on a molecular level. The experiments were done in a physiological saline with 50% extracellular K^+-ions replaced by Cs^+-ions in order to decrease the conductance of K^+-channels. The application of $2.5 \cdot 10^{-5}$ mol/l GABA elicited synaptic current of 2 to 10 nA. The power spectral densities calculated from synaptic current fluctuations could be fitted by single Lorentz curves. The elementary conductance of an inhibitory synaptic chloride channel γ and its mean lifetime τ at a membrane potential E = -100 mV were (T=23°C) : $\gamma = 9 \pm 3$ pS (S.D., n=11), $\tau = 5 \pm 0.7$ ms. Several experiments with physiological saline where K^+-ions were not replaced by Cs^+-ions yielded power spectral densities which could be fitted only by superposition of at least two Lorentz curves. One Lorentz component reflected the usual GABA-induced conductance for Cl^--ions, the other component had a 4-fold lower intensity at 1 Hz and was characterized by a 3 to 4 times higher cut-off frequency. In parallel with the appearance of the faster Lorentz component, in some experiments the total conductance of the fibre, which is normally increased by GABA, was decreased by GABA. This effect could not be suppressed by picrotoxin, indicating that the conductance decrease of the muscle fibre should result from the influence of GABA on non-synaptic membrane channels (Dudel 1979). It seems that this phenomenon involves a decrease of the conductance for K^+-ions (Finger 1979). If this is true the additional Lorentz component is related to the kinetics underlying the closing of K^+-channels.

Dudel, J. (1979) J. Physiol. (Paris), in press.
Dudel, J., Finger, W. and Stettmeier, H. (1977) Neurosci. Letters 6, 203-208.
Finger, W. (1979) Dissertation, Techn. Univ. München, F.R.G.
Neher, E. and Stevens, C.F. (1977) Ann.Rev.Biophys.Bioeng. 6, 345-381.

EXCITATORY SYNAPTIC CURRENT NOISE IN CRAYFISH MUSCLE FIBRES

H. Stettmeier, W. Finger, J. Dudel

Physiologisches Institut der Technischen Universität München

Biedersteiner Strasse 29, 8000 München 40

The application of L-glutamate, the putative excitatory transmitter, to the neuro-muscular junction of the crayfish leads to fluctuations of the membrane current, due to the random opening and closing of ion-selective channels. The statistical analysis of this current noise allows the determination of the mean opentime τ_n and the elementary conductance γ of these channels. In our experiments, single muscle fibres, which had to be very small for good space clamp conditions, were voltage clamped and the membrane currents elicited by superfusion with 50 µmol/l L-glutamate dissolved in physiological saline were recorded. Normally, the currents showed rapid desensitization. Therefore, power spectral densities could be calculated only for the first 5 s after the maximum response and the results from repeated glutamate applications had to be added to improve the statistics of the spectra. In most cases these could be fitted by single Lorentz curves. 14 experiments at a clamp potential E = -60 mV (T=10°C) yielded τ_n = 1.6 $\pm$ 0.3 ms and γ = 21.5 $\pm$ 4 pS while at E = -110 mV (T=10°C) we found τ_n = 1.3 $\pm$ 0.2 ms and γ = 23.5 $\pm$ 5 pS (n=10). The values of τ_n correspond quite well with those found for the decay time of nerve evoked EPSCs under similar conditions. Generally, τ_n was shortened by hyperpolarization though the voltage dependence varied appreciably from fibre to fibre, as also found for the EPSC (Dudel 1979). Increasing the temperature from T=10°C to T=20°C shortened τ_n by a factor of about 0.5.

In several experiments the fibres were pretreated for about one hour with 0.3 µmol/l Concanavalin A (ConA) as this plant lectin blocks desensitization nearly totally (Stettmeier et al. 1978). Under these conditions we found τ_n = 1.8 $\pm$ 0.3 ms and γ = 23.5 $\pm$ 4 pS (n=7) for E = -60 mV (T=10°C). The respective values at E = -110 mV were τ_n = 1.6 $\pm$ 0.4 ms and γ = 22 $\pm$ 7 pS (n=6). Similar changes were observed for the decay of the EPSC. Thus, ConA probably does not affect the elementary conductance, though it may prolong τ_n slightly.

Dudel, J. (1979) J. Physiol. (Paris), in press.
Stettmeier, H., Finger, W., Dudel, J. (1978) Pflügers Arch. 377, R 44/174.

THE SCHWANN CELL SHEATH MODIFIES THE DIELECTRIC PROPERTIES

OF SQUID AXON PREPARATIONS

W. Carius

Chemical Institut of the University of Regensburg

D-8400 Regensburg

The sheath of Schwann cells which envelopes the squid giant axon, is one part of the impedance Z_0 in series with the axon membrane. Other contributions to Z_0 are the electrode and electrolyte resistance inside and outside the axon preparation. In voltage-clamp experiments Z_0 is described as a resistor R_0. Up to 70% of it can be compensated by positive feedback. Experimental data of the membrane admittance are corrected for a series resistance R_0 independent of frequency [1]. The ultrastructure of the Schwann cell sheath [2] has been considered only in calculations on the K^+-accumulation in the periaxonal space [3].

A preliminary report is given on the effect of the Schwann cells on dielectric measurements. A model circuit is proposed taking account of the Schwann cells and the channels crossing them. The model contains distributed electrical circuit elements. At frequencies above 5 kHz it describes well experimental results on the frequency dependance of the admittance reported recently [1]. The dispersion at high frequencies is caused by a capacitive component (C_s) of the series impedance attributed to the Schwann cell layer. This capacitance is far larger than estimated from the contribution of two bilayer membranes surrounding the axon membrane like tubes. This enlargement of the effective membrane capacitance of the Schwann cells is explained by the narrow channels crossing these cells. It is evident from Fig.1, that with a frequency-independent positive feedback loop at voltage-clamp experiments only R_e can be compensated without loss of stability.

Fig.1. Simplified model for the series impedance Z_0

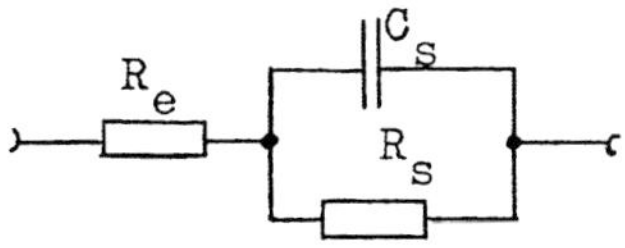

1. Takashima, S., Schwan, H. P.: J. Membrane Biol. 17, 51 (1974)

2. Villegas, G. M., Villegas, R.: J. Gen. Physiol. 51, 44s (1968)

3. Adam, G.: J. Membrane Biol. 13, 353 (1973)

ION-SELECTIVITY OF VOLTAGE- AND OF MECHANO-SENSITIVE MEMBRANE CHANNELS IN STYLONYCHIA

J.E. de Peyer and J.W. Deitmer, Abteilung Biologie, Ruhr-Universität, D-4630 Bochum, West Germany

We have investigated two types of membrane inward currents in the free-swimming ciliate Stylonychia mytilus , which are both largely carried by Ca^{++} ions. One inward current is activated by a mechanical stimulus to the cell anterior (mechano-receptor current), the second one is activated by depolarizing voltage pulses (voltage-dependent current). We have studied these membrane inward currents during voltage clamp in the presence of Ca^{++}, and when Ca^{++} was replaced by Mg^{++} in the bathing solution. The voltage-dependent inward current fully disappeared in the Ca^{++}-free, Mg^{++}-containing solution, while a mechanical stimulus still evoked an inward current in this solution. The reversal potential of this receptor current varied with the extracellular concentration of both Ca^{++} and Mg^{++} (the slope in a semi-log plot was 23.5 mV and 21.0 mV respectively). We have tested all other ions present in the bathing solution (K^+, Cl^- and $Tris^+$) whether they contribute to the mechanoreceptor current: only a small (outward) current carried by K^+ ions appeared to be involved in this current. Our results indicate that mechano-sensitive and voltage-sensitive channels in Stylonychia have a different selectivity for Ca^{++} and Mg^{++} ions: the voltage-sensitive channel is permeable to Ca^{++} but not to Mg^{++} ions, while the mechano-sensitive channel is permeable to both ions, and does not discriminate between them.

In the absence of extracellular Ca^{++} ions the delayed, voltage-dependent outward current (outward rectification of the membrane) is greatly reduced or even abolished, suggesting that this rectification depends on extracellular Ca^{++} and /or on the influx of Ca^{++} into the cell. Coupling between the mechanical stimulus and the ciliary motor response appears to depend directly on the influx of Ca^{++} ions through the voltage-sensitive channels.

Supported by the Deutsche Forschungsgemeinschaft, SFB 114 (TP A5).

Functional separation of subcellular sensitive elements by selective adaptation in a ciliary mechanoreceptor cell.

J. Gödde and U. Thurm

Lehrstuhl für Neurophysiologie, Zoologisches Institut
Hüfferstr. 1,D-4400 Münster

Single mechanoreceptor cells associated with large tibial hairs of crickets can be stimulated by differently distributed deformations obtained by bending the hair in different directions. The cylindrical sensitive cell region, i.e. the apical 1.5 µm of a ciliary outer segment, is stimulated by an indentation, executed by the hair base, in any of its perpendicular directions. The adaptation of the receptor potential response proves to be associated with the particular distribution of the deformation which elicited the response. The cell-physiological nature of this adaptation (in contrast to a purely mechanical one) is indicated by, among other observations, a marked dependence of the rate of adaptation on transepithelial voltage. We use the selectivity of this adaptation to get information on distribution and properties of the subcellular mechanosensor elements, since the distinction at the level of adaptation to differently distributed deformations strongly indicates a difference in the effect of these stimuli already at the preceding level of molecular sensors.

The receptor potential versus the degree of bending characteristic is similar for all directions of bending. Adaptation results in a parallel shift of stimulus-response curves along the stimulus axes towards higher stimuli.

The adaptation achieved by a conditioning stimulus is found to diminish a test response less the more the direction of the test stimulus deviates from the direction of conditioning. The minimum of adaptation is found for responses elicited by stimuli which are applied in the quarter opposite to the direction of conditioning. This is valid for all directions of conditioning.

A fully adapted response can be continued directly by a much higher, little adapted response amplitude if the bending angle of the hair is kept constant but the direction of bending is subsequently rotated. As long as this rotation is continued the response amplitude is considerably increased towards the unadapted (phasic) amplitude.

This, and other results obtained by partial rotations, indicate that the total sensitivity of the receptor ending can be best explained by a continuously distributed population of sensor elements of which only overlapping subpopulations are stimulated during differently directed bendings of the hair.

OPTICAL MONITORING OF MEMBRANE POTENTIAL CHANGES AS A TOOL TO STUDY SODIUM-
DEPENDENT SUBSTRATE TRANSPORT IN BRUSH BORDER VESICLES FROM KIDNEY PROXIMAL
TUBULES.

G. Burckhardt, H. Murer

Max-Planck-Institut für Biophysik, 6000 Frankfurt/Main

Sugars and neutral amino acids are accumulated in epithelial cells of small
intestine and renal proximal tubule. The uptake of these substrates across the
luminal cell membrane is coupled to sodium and is potential sensitive. Changes
in the membrane potential difference in vesicles derived from rabbit kidney
proximal tubule can be monitored by a permeant cyanine dye, 3,3'-diethylthia-
dicarbocyanine iodide. This positively charged dye responds within a few seconds
to changes of the potential difference. The addition of D-glucose to Na_2SO_4
equilibrated vesicles causes a concentration dependent transient fluorescence
increase indicating a depolarization. L-alanine and L-phenylalanine cause a
fluorescence increase which lasts a shorter time than glucose-induced changes
indicating a faster equilibration of these amino acids across the plasma membrane. Fluorescence changes are not detected if Na^+ is replaced by K^+ or if
gramicidine is added to "short circuit" membrane potential changes.

The simultaneous addition of saturating concentrations of D-glucose and
L-alanine leads to a partial addition of the individual fluorescence changes.
This is taken to be indicative for two independent transport systems. In
contrast, if at saturating substrate concentration a substrate is added,
which is transported via the same system, no further change in fluorescence
is obtained. By this method it might be possible to test the specificity of
amino acid transport systems.

Attempts are made to find a relationship between fluorescence and membrane
potential difference. In the presence of valinomycin successive replacement
of K_2SO_4 by $(choline)_2SO_4$ in the medium leads to a gradual fluorescence
decrease with K_2SO_4 loaded vesicles (increased inside negativity). Under
these conditions, the membrane potential should tend towards a potassium
diffusion potential. The fluorescence is a S-shaped function of the membrane
potential showing a saturation at extreme inside positive and inside negative
membrane potentials.

IS THE SPONTANEOUS LEAK FOUND IN THE RESEALED GHOST RELATED TO THE GARDOS EFFECT?

P.G.Wood and U. Rempel-Rossleben, Max-Planck Institute for
Biophysics, D-6000 Frankfurt am Main, FRG

In the intact human red blood cell, the passive leak for
potassium is very low. In contrast the resealed ghost can
display a spontaneous leak for potassium, when hemolysis
is conducted in the absence of magnesium at high dilution
of cell contents. The magnitude of the leak is reduced when
chelators or magnesium are incorporated at hemolysis. Alternatively, the spontaneous leak is inhibited when the resealed
ghosts are washed in solutions containing chelating agents
such as EDTA, EGTA, and ATP. Thus, metal ion binding sites
at the inner and outer surface of the membrane can influence
the state of the potassium barrier.

In the Gardos Effect metabolic depletion and the elevation
of internal calcium is required to induce the potassium leak.
The spontaneous leak may be due the depletion of endogenous
chelators and the release of calcium from the membrane at
hemolysis. To test this hypothesis, terbium and europium
have been used as probes of calcium. When either is applied
to the inner surface through incorporation at hemolysis(0.1-
10 μM), the magnitude of the potassium leak in the resealed
ghost markedly increases. However, when they are applied
in excess to the outer surface(20 μM) after resealing, both
the spontaneous potassium leak and the leak generated by
internal terbium, europium, or elevated internal calcium
is immediately inhibited. The state of the potassium barrier
is, therefore, controlled by at least two metal ion binding
sites which act in opposition.

Furthermore, the net exchange of internal potassium into
choline medium is stimulated by low concentrations of external
potassium in both the spontaneous leak and that of the Gardos
Effect. External terbium and europium interact with external
potassium to inhibit the spontaneous leak. In contrast the
pH dependence of the two leak systems is different. In
general, however, the spontaneous leak does seen to be related
to that of the Gardos Effect. The differences observed may
be more due to concentration levels of calcium that appear
at the inner surface, rather than to fundamental differences
between the two pathways.

TRANSFORMED CELLS DIFFER FROM NORMAL CELLS BY THE TOPOLOGICAL
ARRANGEMENT OF THE OUTER CELL MEMBRANE

J.P. Seher and G. Adam

Fakultät für Biologie, Universität Konstanz, D-7750 Konstanz

The geometrical surface area of 3T3 cells grown in vitro is at least by a factor of 2.5 larger than that of their SV40-virustransformed derivatives [1,2], independent of cell growth density [2]. In the present study, we have characterized cell surface areas of normal and transformed 3T3 cells at low cell densities additionally by the following independent experimental methods. An electrically effective surface area A_E was determined electrophysiologically as the ratio r_M/R_M of specific membrane resistance r_M and total membrane resistance R_M, assuming a specific membrane capacity of 1 μF cm^{-2}. Geometrical cell surface areas A_S were determined after severe swelling, which may be achieved by a procedure of superficial drying of cells [3]. Geometric cell surface areas were evaluated by microscopic calibration of cells by high-resolution differential interference contrast: surface areas O referring to cells attached to the growth substrate and O_S referring to suspended cells exposed to strongly hypotonic medium. Furthermore, cell surfaces were characterized by labeling primary amines of the outer membrane of cells with the short-lived reagent fluorescamine [4] and evaluating the ratio of relative fluorescence of 3T3 cells and SV40-3T3 cells, where F refers to cells on the growth substrate and F_d refers to suspended cells labeled after sonication. Experimental surface areas in 10^{-5} cm^2 ± S.E.M. of 3-5 experiments and ratios of relative fluorescence are given in the following table, all data referring to a cell density of $3 \cdot 10^3$ cm^{-2}, except for F_d which refers to $8 \cdot 10^3$ cm^{-2}.

	A_E	A_S	O	O_S	F	F_d
3T3	4.7 ± 0.8	4.4 ± 0.4	5.4 ± 0.6	-	2.16	1.08
SV40-3T3	5.7 ± 1.0´	4.2 ± 0.6	2.2 ± 0.2	5.7 ± 0.5	(1)	(1)
PY-3T3	-	4.8 ± 0.5	-	-	1.19	1.05

Our interpretation of these data is that growing transformed cells have involuted a larger part of their total surface membrane and thus rendered it virtually inaccessible to the surface reagent fluorescamine within its (short) effective time of reaction. The total surface membrane area of transformed cells, however, is electrically effective (i.e. passively permeable to ions) and can be exposed by severe swelling. It turns out to be equal within experimental error to that of normal cells.

[1] Collard, J.G. and Temmink, J.H.M., J.Cell Sci. 19 (1975) 21-32.
[2] Seher, J.-P. and Adam, G., Z.Naturforsch. 33c (1978) 739-743.
[3] Adam, G. and Schumann, C., Prog. Colloid and Polym. Sci. 65 (1978) 200-205.
[4] Hawkes, S.P., Meehan, T.D. and Bissele, M.J., Biochem.Biophys.Res. Commun. 68 (1976) 1226-1233.

Supported by DFG/SFB 138

FLUORESCENCE POLARIZATION OF 3T3 AND SV40-3T3 CELLS THINLY SPREAD ON THEIR GROWTH SUBSTRATE

U. Steiner and G. Adam

Fakultät für Biologie, Universität Konstanz, D-7750 Konstanz

Evaluation of fluorescence polarization in an experimental arrangement of spherical symmetry misses some of the information available from experiments with lesser symmetry [1]. Furthermore, the state of cells thinly spread on their growth substrate may be different from that of suspended cells. We have, therefore, investigated polarization of fluorescence of 1,6-diphenyl-1,3,5-hexatriene (DPH) incorporated into cells grown on, and still attached to, microscope cover slips. Experiments were carried out on a Perkin-Elmer MPF-44A fluorescence spectrometer equipped with polarizers. The cover slips were inserted diagonally into a normal 10 mm quartz cuvette, so that the angle of incidence of the exciting beam to the plane growth surface was 45° with the electric vector parallel to it.

Experimental results differed from those on suspended cells [2,3]. The degree of polarization $P = (I_{\|} - I_{\perp})/(I_{\|} + I_{\perp})$, where $I_{\|}$ and $I_{\perp}$ are intensities of fluorescent light measured with the analysator parallel or vertical, respectively, to the polarizer, decreases for both cell lines by about 20% within the first 30 min of incubation with the DPH-containing solution and remains essentially constant thereafter.

Measured after 20 min of incubation with DPH label, the degree of polarization for both cell types was found equal within experimental error: $P = 0.15 \pm 0.01$.

Comparing these results with those on suspended cells [2,3], it is concluded that the physical state of the lipid/protein matrix of cellular membranes and/or the geometrical arrangement of different portions of cellular membranes are altered differently for normal and transformed 3T3 cells upon suspension.

[1] Frehland, E., Z.Naturforsch. 30a (1975) 1241.
[2] Fuchs, P., Parola, A., Robbins, P.W. and Blout, E.R., Proc.Natl. Acad.Sci. USA 72 (1975) 3351.
[3] Inbar, M., Yuli, I. and Raz, A., Exp. Cell Res. 105 (1977) 325.

Supported by DFG/SFB 138

THE EFFECT OF BLEOMYCIN ON V 79-SPHEROIDS

Tobüren, D.

Institut für Strahlenbiologie

Universität Münster

Spheroids resemble solid tumors in vivo not only in regard to cell kinetics but also to nutrition and oxygen supply of the cells they consist of. Therefore, spheroids provide to be a useful in-vitro tumor model owing the complex architecture of solid tumors in vivo but still retaining the advantages of in vitro culture methods. The purpose of the experiments was to investigate the effect of Bleomycin (BLM) on spheroids and their different subpopulations in comparison to cells growing in petri-dishes, i.e. monolayers, at different growth phases (log-phase, fed and unfed plateau-phase).

Dose-response curves obtained by incubation with BLM up to loo μg/ml for one hour are biphasic with a sensitive part at low doses and a resistant part at higher doses, no matter whether the cells were grown as monolayers or as spheroids. Spheroid-cells are slightly more responsive to the drug at concentrations higher than 5 μg/ml than any monolayer-culture tested. Furthermore, prolonged treatment with BLM up to 24 hours leads to an augmented killing effect of spheroid-cells in comparison to exponentially growing cells in petri-dishes, the most sensitive monolayer-cultures. Cells originating from the outer, intermediate and inner zone of the spheroids exhibit the same sensitivity when treated for one hour. In Addition, elimination of hypoxia within the spheroids by lowering the incubation-temperature, results in about the same survival-values of log-phase monolayer cultures, spheroids and of cells from the two inner zones of spheroids. Therefore, the influence of hypoxia may be less important to the differential surviving ability of the spheroid-subpopulations which is observed after prolonged treatment at 37°C.

DISTURBANCE OF PROLIFERATION KINETICS OF L-CELLS BY ^{3}H-THYMIDINE LABELLING

Hans-Peter Beck

Institut für Biophysik und Strahlenbiologie

Universität Hamburg

Martinistr. 52, 2000 Hamburg 20

In autoradiographic studies the decay of the radioactive isotopes incorporated into the DNA leads to disturbed proliferation kinetics of the labelled cells even at low concentrations of activity (1,2). In the present paper these cell kinetic parameters were determined for a L-929 cell population in vitro after flash labelling with ^{3}H-thymidine (30 minutes, 0.3 µCi/ml, 40 Ci/mM) using the flow cytometric BUdR-33258 technique (3).

The histogram series obtained were evaluated using a new procedure that allows to calculate the growth curve of the cells during the experiment and to normalize the histograms of the series. From the normalized histograms, the influx into the G2 compartment and the eflux from G2 was calculated in order to determine the behaviour of subpopulations of cells during their transit through G2.

The results show that one fraction of the labelled cells is delayed in the S phase but transits through G2 with normal velocity, whereas another fraction is not delayed in S but is considerably retarded in the G2 compartment. The results also show the degree of radiosensitization of the cells that is due to BUdR incorporation.

(1) Beck, H.-P. (1978) Störungen des Zellzyklus durch radiotoxische Effecte. Poster auf der Jahrestagung der Deutschen Gesellschaft für Biophysik.
(2) Beck, H.-P. and M. Omniczynski (1979) Comparison of the applicability of flow cytometry and autoradiography for cell kinetic studies. Radiotoxic effects of incorporated 3H-thymidine and tumour response to irradiation. 4. International Symposium on Flow Cytometry, in press.
(3) Böhmer, R.M. (1979) Flow cytometric cell cycle analysis using the quenching of 33258 hoechst fluorescence by bromodeoxyuridine incorporation. Cell Tissue Kinet. <u>12</u>, 101.

Scanning Electron Microscopic Measurement Of Human Red Blood Cell Diameter Alterations As A Function Of Cell Age

W. Strobelt [+], Th. Vömel [++] and D. Platt [++]

+) Institut für Biophysik der Universität Giessen
   Strahlenzentrum, Leihgesterner Weg 217, D-6300 Giessen
++) Institut für Gerontologie der Universität Erlangen-Nürnberg
   Flurstraße 17, D-8500 Nürnberg

The aging process in the normal human erythrocyte is accompanied by a variety of biochemical and biophysical alterations. Previous studies have discribed differences in the red cell volume dependent upon aging, but with contrary results (1)(2)(3). In these investigations Coulter electronic particle counting systems were used for volume determination. The different results may be attributed to artifacts of the Coulter counter technics. So it is obvious to try the scanning electron microscopy (S 4 Stereoscan, Cambridge Scientific Instruments, England) as another method to prove the size of red blood cells in correlation with the age of the cells.

Altogether we measured in 14 healthy persons diameters of 6857 erythrocytes after separation by ultracentrifugation (4) into three groups of different age, 2269 younger cells, 2254 of middle age and 2334 older ones. We got for the mean value of the younger erythrocyte diameters 7,8 μm, of the middle aged cells 7,3 μm and of the older cells 7,0 μm. The normal range of the values in the younger group was 8,0 to 7,2 μm, in the middle aged 7,8 to 7,0 μm and in the older Group 7,5 to 6,7 μm.

The results suggest a decrease of the erythrocyte diameter during the aging process.

(1) Coopersmith, A. and M. Ingram:
    Am. J. Physiol. 215/5, (1968), 1276-1283
(2) Stenhouse, N. and H.J. Woodriff:
    Med. J. Australia 19,(1974), 614
(3) Ganzoni, A.M., J.P. Barras and H.R. Marti:
    Vox Sang. 30, (1976), 161-174
(4) Platt, D., K.H. Hofmann and T. Vömel:
    Z. Gerontologie 12, (1979), 60-72

INFLUENCE OF REDUCING AND OXIDIZING SUBSTANCES ON HEALTHY AND "LEUKEMIC" BLOOD AND ITS FRACTIONS

J. Schreiber, W. Greulich, and W. Lohmann
Institut für Biophysik der Justus-Liebig-Universität
Giessen, Strahlenzentrum,
Leihgesterner Weg 217
6300 Giessen, FRG

Lyophilized healthy blood as well as erythrozytes exhibit an electron
spin resonance (ESR) signal at about $g = 2.005$. Blood samples of
patients with leukemia have an additional peak and an increased spin
concentration. In order to elucidate the possible causative agents
as well as their prospective receptors, several substances have been
added to either native blood, erythrocytes and their white ghosts,
leukocytes or plasma of healthy persons. Of all the substances tested,
only ascorbid aced produces a spectrum which is identical to the one
obtained from blood or its fractions of "leukemic" patients. Since
these changes were observed in both white ghosts and plasma, the
receptor for vitamin C had to be searched for in membrane and plasma
as well. The interaction between ascorbid acid and transsition metal
ions, being present in these blood constituents as revealed by atomic
absorption studies, didn't result in the typical "leukemic" signal.
The interaction between lyophilized samples of ascorbic acid and some
copper-containing proteins (ceruloplasmin, cytochrom-c-oxidase, haemocyanin) resulted in spectra which are identical to the one obtained
with "leukemic" blood. A model will be proposed which can explain the
experimental findings reported thus far (e.g. change in spin concentration with the development of cancer, the presence of a high concentration of antioxidants etc.). The results obtained suggest that
ascorbic acid and its possible interaction with copper-containing
proteins might play an important role in the case of leukemia.

This work was supported in part by Euratom grant EUR no.
213-76-7 BIO D.

INTERACTION BETWEEN ASCORBIC ACID AND Cu-PROTEINS: ATOMIC ABSORPTION AND ESR MEASUREMENTS ON ERYTHROCYTE GHOSTS AND BLOOD PLASMA

W. Greulich, J. Schreiber, and W. Lohmann
Institut für Biophysik der Justus-Liebig-Universität
Giessen, Strahlenzentrum,
Leihgesterner Weg 217
6300 Giessen, FRG

Lyophilized samples of mixtures of healthy native blood or erythrocytes with ascorbic acid exhibit an electron spin resonance (ESR) signal which has been obtained with leukemic blood. In order to find out whether membrane constituents are involved in the ascorbic acid interaction, white ghosts of erythrocytes were prepared according to the method of Dodge et al. (Arch. Biochem. Biophys. 100 (1963) 119). Lyophilized white ghosts which contained $\leq$ 0.1% of the erythrocyte hemoglobin didn't show any ESR signal. Addition of ascorbic acid resulted in the typical "leukemic" ESR signal. Since similar changes were also observed in the plasma the receptor of vitamin C has to be searched for, therefore, in the membrane and plasma as well. It should be pointed out that the additional peak caused be the interaction with vitamin C is already present in plasma of healthy persons. In the case of leukemic patients the spin concentration is, however, reduced considerably. Based on previous findings, metallo proteins might function as receptors for vitamin C. Samples of ascorbic acid and catalase or cytochrome c didn't reveal the ESR spectrum of leukemic blood. Copper containing proteins (e.g. ceruloplasmin, cytochrom-c-oxidase, haemocyanin), however, exhibit the "leukemic" signal when mixed with ascorbic acid. From this, it might be concluded that copper must be present in the membrane. Atomic absorption studies have shown that Cu is present in white ghosts (4×10^4 Cu atoms/WG). Furthermore, the copper content of leukemic plasma is about twice of that of healthy one, while the Fe content is diminished. Thus, the signal observed in leukemic blood seems to be due to a copperprotein-ascorbic acid interaction. In the case of leukemia, the vitamin C metabolism seems to be modified.

This work was supported by Euratom grant EUR no. 213-76-7 BIO D.

ORGAN SPECIFIC APPLICATION OF DRUGS BY MEANS OF CELLULAR CAPSULE SYSTEMS

G. Pilwat, J. Vienken, U. Zimmermann
Institute of Chemistry (Biophysical Chemistry)
Nuclear Research Center Jülich
P.O.Box 1913, 5170 Jülich, W.-Germany

Abstract

It is suggested to use living cells (red blood cells, lymphocytes and leucocytes) as drug delivery systems for temporal and spatial drug administration in human therapeutics and diagnosis. The effectiveness of drug loaded cells is demonstrated for the drug methotrexate which is used in cancer treatment.

Red blood cells are loaded with methotrexate using the dielectric breakdown technique. Dielectric breakdown leads to a transient increase of permeability of the cell membrane. Red blood cells loaded with tritium-labelled methotrexate were injected into mice and the activity level was measured in several organs as a function of time. It is shown that with this drug delivery system more than 50% of the drug (after 10 min) can be accumulated in the liver and that a high activity level can be sustained in this organ for more than 3 hours. On the other hand, administration of this drug by injecting solutions in the usual manner leads only to an 25% accumulation of methotrexate (after 10 min) in the liver. The drug is excreted completely after 1 to 2 hours.

It is proposed to load red blood cells simultaneously with para- or ferromagnetic substances to obtain organ-specificity for any selected site of the body.

Lymphocytes can also be used for directing drugs to selected sites of the organism whereby the lymphocyte carrier system is of considerable interest to overcome the brain-blood barrier and to target drugs to the brain.
It is demonstrated that lymphocytes can be loaded electrically without loss of cell membrane integrity and cellular functions.

INVESTIGATION OF BIOPHYSICAL PARAMETERS IN
DEEP FREEZING EXPERIMENTS WITH PLANT CELLS
IN ORDER TO ACHIEVE A HIGH RATE OF SURVIVING
FROST SENSITIVE CELLS

K.-H. C. Standke

Molekularbiologisches u. Biophysikalisches Laboratorium,
D-3300 Braunschweig, FRG and
Lehrstuhl A und Institut für Physik, Chemie u. Elektro-
chemie der Techn. Universität Braunschweig,
D-3300 Braunschweig, FRG

In 1973 K. K. Nag and H. E. Street (1) obtained a surviving rate of
67 % of single carrot cells after deep freezing at -196° C. Meanwhile,
Lindsay A. Withers (2), B. W. Grout and G. G. Henshaw (3) as well as
K.-H. C. Standke (4) successfully employed -196° C for deep freezing
of somatic embryos of carrots and meristems from shoot tips of non-
frost resistant potatoes (solanum tuberosum). It turned out from these
experiments that survival of meristems from shoot tips of potatoes
requires an adequate exposure of plant material to the conditions of
minimum growth at temperatures between +4° C and +10° C before plunging
into liquid nitrogen. With this method the surviving rate of the
meristems ranges between only 10 and 20 % due to the fact that relatively far developed material has been employed.

A increased surviving rate of constantly 40 % is achievable when from
potatoe shoot tips cv. "Grata" apical domes with only 2 leaf primordia
are beeing used; otherwise, any material with further differentiated
cells very easily tends to die in contact with liquid nitrogen.

From these experiments it is obvious that a very high surviving rate
of meristems and embryos even from non-frost resistant plants could
be expected. In this connection it is of vital importance that the
state of liquid water be maintained. For this reason a shock freezing
cooling rate of at least 500° C/min is required.

(1) Nag, K.K. and Street, H.H., Nature, 245, 270 (1973).
(2) Withers, Lyndsey A., Plant Physiol., 63, 460 (1979).
(3) Grout, B.W.W. and Henshaw, G.G., Ann. Bot. 42, 1227 (1978).
(4) Standke, K.-H.C., Hoppe Seyler's Zeitschrift f. Physiol. Chemie,
    357, 283 (1976).

NATURAL ELECTRIC CURRENTS TRAVERSE GROWING PLANT CELLS AND TISSUES.

M.H. Weisenseel, Institut für Botanik und Pharmazeutische Biologie der Universität Erlangen-Nürnberg

The existence, pattern and density of natural currents traversing developing plant cells and tissues have been investigated with the aid of a ultrasensitive vibrating electrode that can detect such currents extracellularly with a spatial resolution of 20-30 μm. Radioisotopes and ionspecific dyes have been employed to elucidate the ionic components of these currents.

In all plant cells and tissues investigated so far, i.e. in the zygotes of the brown algae Fucus and Pelvetia (1,2), in the pollen grains and pollen tubes of Lilium (3,4), in the root hairs and roots of Hordeum (5) and in the green alga Vaucheria (6) currents of a few $\mu A\ cm^{-2}$ always enter the site of future or of actual growth and leave the non-growing part of the cell or tissue. The ions that carry these currents are Na^+, Ca^{2+} and Cl^- in the brown algae (Na^+ and Ca^{2+} enter the rhizoid end and Cl^- enters the thallus end); K^+, Ca^{2+} and H^+ in lily pollen (K^+ and Ca^{2+} enter the pollen tube, H^+ ions are pumped out at the pollen grain) and H^+ and Ca^{2+} in root hairs (H^+ and Ca^{2+} enter the growing tip, H^+ ions leave the root surface). H^+-ions enter the main elongation zone of the barley root and leave the surface of the root-hair-zone.

Since these selfgenerated currents always <u>precede</u> actual growth they seem to be an essential step in the process of establishment of localiced growth and not just a byproduct of this growth. They might become effective by inducing local differences in ion concentrations and establishing electric fields within the cytoplasm and along the cell membrane.

1) Jaffe, L.F., Robinson, K.R., Nuccitelli, R. In: Membrane Transport in Plants. Eds. U. Zimmermann and J. Dainty, 226-223 Springer (1974) -
2) Nuccitelli, R.: Dev.Biol. 62, 13-39 (1978) -
3) Weisenseel, M.H., Nuccitelli, R., Jaffe, L.F.: J.Cell Biol. 66, 556-567 (1975) -
4) Weisenseel, M.H., Jaffe, L.F.: Planta 133, 1-7 (1976) -
5) Weisenseel, M.H., Dorn, A., Jaffe, L.F.: Plant Physiol. in press -
6) Weisenseel, M.H.: work in progress

SELECTIVE ACCUMULATION OF ALKALI-METAL IONS AT CELLULAR PROTEIN SITES WITHOUT ION PUMPS

Ludwig Edelmann

Medizinische Biologie, Universität des Saarlandes, D-6650 Homburg

The membrane theory maintains that the high K^+ level inside a cell is dependent on a functioning Na/K pump and that the cell K^+ is dissolved in the cell water. On the other hand, the association-induction hypothesis (AIH) maintains that the bulk of cell K^+ is selectively adsorbed to fixed anions of cell proteins.[1] Recent experimental findings strongly support this hypothesis: a) An effectively membrane-pumpless open-ended cell preparation continues to demonstrate K^+ accumulation and Na^+ exclusion much as a normal cell does.[2] Independent measurements confirm the prediction of the AIH that cellular K^+ (Rb^+, Cs^+) accumulation occurs at proteins carrying much β-and γ-carboxyl groups (myosin).[3,4] Further evidence for the idea that selective K^+ accumulation might be due to binding onto cellular proteins is provided by the following experiments: Frog sartorius muscles were freeze-dried, embedded and sectioned. Sections 0.1-0.2 μm thick were exposed to aqueous solutions containing different concentrations of LiCl, NaCl, KCl and CsCl. Energy dispersive X-ray microanalysis (Siemens, ST 100 F) and laser microprobe mass analysis (Leybold Heraeus, LAMMA 500) of these sections yielded the following consistent results:

1) The sections accumulate each species of alkali-metal ions; the ions are preferably localized in the A band (myosin).

2) Selective accumulation of the alkali-metal ions is detectable if the sections are exposed to 2 or more species of the alkali-metal ions.

3) If the sections are exposed to equal concentrations of K^+ and Na^+ (e.g. 10 mM) much more K^+ than Na^+ can be detected in the A band; the concentration of accumulated K^+ is in the same order of magnitude as the K^+ concentration of a living cell.

4) If the sections are exposed to Na^+, K^+ and Cs^+ the rank order of selectivity is $K^+ > Cs^+ > Na^+$; if the sections are exposed to Li^+, Na^+, K^+ and Cs^+ the rank order of selectivity is $Li^+ > Cs^+ \gg K^+ > Na^+$. (Due to the high Cs^+ accumulation the sections are well stained.)

2) and 3) show that myosin fixed in the natural 3-dimensional configuration can selectively accumulate K^+ and that hypothetical ion pumps are not necessary to explain the high K^+ content of living cells. The change of the selectivity order of Na^+, K^+ and Cs^+ in the presence of Li^+ can be explained by an indirect F-effect (see (1) chap. 5).

1. G.N.Ling, A Physical Theory of the Living State. Blaisdell (1962)
2. G.N.Ling, J. Physiol. 280, 105 (1978)
3. L.Edelmann, Microsc. Acta, Suppl.2, 166 (1978)
4. L.Edelmann, Physiol.Chem.Phys. 9, 313 (1977) and 10, 469 (1978)

Section C

NEUROBIOLOGY AND CYBERNETICS

EXCHANGE OF SOLUTES BETWEEN SUPERFUSATE AND THE EXTRACELLULAR COMPART-
MENT ADJACENT TO THE PHOTOSENSORY MEMBRANE OF *LIMULUS* AND CRAYFISH PHOTO-
RECEPTOR

H. Stieve, W. Schröder, M. Bruns, I. Claßen-Linke
Institut für Neurobiologie, KFA Jülich, Postfach 1913
D-5170 Jülich/ F R Germany

1a) In the *Limulus* ventral nerve preparation addition of 1 mmol/1 La^{+++}
to the superfusate does not reduce the electrical light response (receptor potential) to a bright constant stimulus.

   In the same *living* preparation neither La^{+++} - determined by
Laser-Micro-Mass-Spectrometry (LAMMA) - nor micro-peroxidase (MP) (M.
W. ca. 1600) - as shown by light- and electronmicroscopy (LM and EM)
- reaches the microvillar membrane even when the preparation was previously treated with pronase in a standard procedure used for electrophysiological experiments. However, if applied after fixation of the
ventral nerve with aldehydes, lanthanum reaches the extracellular
space of the microvillar region.

1b) In the isolated *Astacus* retina preparation addition of 1 mmol/1 La^{+++}
to the superfusate abolishes the electrical light response (ERG).

   In the same *living* preparation La^{+++} or MP is able to reach the
small extracellular spaces between the photosensory microvilli of the
rhabdome (LAMMA, LM, EM).

2a) Lowering the Ca^{++} and Mg^{++} concentration in the superfusate to very
low values (< 10 µmol/1 by adding 1 mmol/1 EDTA) abolishes in *Limulus*
ventral nerve photoreceptors the electrical light response.

2b) In the *Astacus* retina preparation reduction of the Ca^{++} and Mg^{++} concentration in the superfusate to < 100 nmol/1, using 1 mmol/1 EDTA,
causes only a small reduction in light response. A much stronger and
faster diminution of the light response is caused when 10 mmol/1 EDTA
is added to the superfusate without calcium and magnesium.

Interpretation: Solutes can reach the photosensory membrane in the *Astacus*
retina preparation by diffusion via solely extracellular pathways whereas
in the *Limulus* photoreceptor preparation solutes have to pass a diffusion
barrier which allows Ca^{++} but not La^{+++} or MP molecules to penetrate.
In the *Astacus* retina preparation the extracellular microvillar space in
the rhabdome is difficult to deplete of Ca^{++}, since this remote compartment is refilled with Ca^{++} by intracellular calcium sources.

Supported by DFG (SFB 160)

LASER-MICROPROBE-MASS-SPECTROMETRY REVEALS TWO CLASSES OF SHIELDING PIGMENT IN THE *ASTACUS* RETINA: ONE BINDING LARGE AMOUNTS OF CA, THE OTHER BINDING NA AND K.

W. Schröder[1], D. Frings[1], and H. J. Heinen[2]

[1] Institut für Neurobiologie der Kernforschungsanlage Jülich GmbH,
   Postfach 1913, D-5170 Jülich
[2] Leybold-Heraeus GmbH, Gaedestr., D-5000 Köln 51

Following the question: are there intracellular stores for Na, K, Ca in the retina of evertebrates? We analysed the distribution of these elements with a laser-micro-mass-analyser (LAMMA). Sections of some 0.5 μm were examined with a conventional phase contrast light microscope fitted to a time of flight mass spectrometer. A small spot of the section was evaporated and ionized by a focussed UV power-pulse-laser. The spatial resolution is better than 0.5 μm, the sensitivity of the mass-spectrometer lower than 3000 atoms of Ca.

Line scans through the retina showed only one structure associated with larger amounts of Ca: the distal shielding pigment granules (DP). A rough estimation suggests a concentration of some 100 to 150 mmol/l Ca in or associated with this type of pigment. Lower concentrations of Ca are found within the rhabdome (some 5 mmol/l) or at the intracellular locations closely adjacent to the distal pigment (less than 1 mmol/l).

Analysis of the proximal shielding pigment granules (PP) revealed high values for Na and K (from 50 to 200 mmol/l) but very low values for Ca (less than 0.1 mmol/l).

Incubation of retinas in solutions of 1 mmol/l EDTA quickly lowers the Ca concentration within the area of the basal membrane to less than 1 μM but does not drastically lower the Ca content of the DP within 30 min. Incubation in 50 mmol/l Ca^{++} for several hours results in a large increase of Ca in the DP. However only traces of Ca can be found in the PP under these conditions.

Our interpretation is that the DP acts as a Ca store, the PP as a store for Na and K possibly regulating the concentration of these ions.

Supported by DFG/SFB 160

LIGHT-INDUCED CALCIUM RELEASE FROM BOVINE DISK VESICLES

Paul J. Bauer, and H. Gilbert Smith, Jr.[+]
Institut f. Neurobiologie, KFA Jülich, Postf. 1913, D-5170 Jülich
[+]Current address: Dept. of Physiology and Biophysics, Washington
University, St. Louis, Missouri 63110

Light-induced calcium permeability changes of disk membranes were
studied using the metallochromic calcium indicator, arsenazo III.
Calcium containing disk vesicles were prepared by sonication of
purified disks which were obtained from bovine rod outer segments
by Ficoll flotation (Smith et al. (1975), Exp. Eye Res. 20, 211).
The vesicles were allowed to reseal overnight at 4 °C. Then, the
external calcium was removed by treatment with the ion exchange
resin, Chelex 100, and arsenazo III was added. Absorbance changes
were monitored at 652 nm. The following results were obtained:
1. light induces a transient calcium permeability change leading
to a partial release of the internal calcium;
2. the permeability increase subsides with a time constant in the
range of minutes at 37 °C;
3. addition of the calcium ionophore, A23187, results in a release
of the remaining internal calcium;
4. the magnitude of the light-induced calcium release increases with
both the amount of internal calcium and with the amount of rhodopsin
bleached;
5. the magnitude and the kinetics of the calcium release is
temperature dependent;
6. the light-induced calcium releases were normally found to be
greater than one calcium per bleached rhodopsin at 37 °C;
7. the amount of calcium released per bleached rhodopsin increases
with decreasing percent bleaching;
8. the action spectrum of the calcium release parallels the
absorption spectrum of rhodopsin.
These results strongly indicate a transient calcium permeability
increase of disk membranes after light absorption. The specificity
of this permeability change was studied with radioactive substances
using a flow system technique (Smith et al. (1977), Biochemistry 16,
1399). Definite light induced releases of variable magnitude were
found for ^{86}Rb, ^{22}Na, ^{32}P-phosphate, ^{3}H-glucose, and ^{3}H-sucrose
indicating that the permeability change is relatively non-specific.

THE KINETIC BEHAVIOUR OF RAPID CALCIUM RELEASE IN CATTLE ROD OUTER SEGMENTS:THE METARHODOPSIN I/METARHODOPSIN II-TRANSITION IS INVOLVED IN THE TRIGGER MECHANISM

U.B.Kaupp,P.P.M.Schnetkamp and W.Junge [§]

Max-Volmer-Institut,Techn.Univ.Berlin,1 Berlin 12,Strasse des 17.Juni 135,Germany

Only recently a rapid,light-induced calcium release in intact ROS resulting from a change of the stability constant of calcium binding sites at the disc membrane has been reported(Kaupp et al.(1979)Biochim.Biophys.Acta 552,39o-4o3).By applying kinetic flash spectrophotometry with the calcium-indicating dye arsenazo III the kinetic and stoichiometric behaviour of the release have been further investigated.

   Upon fragmentation of intact ROS by lysis or sonication,the light effect on the calcium binding sites disappears.This is due to a difference in accessibility of the binding sites for monovalent cations in intact ROS and fragmented material.Intact ROS are a three compartment system with an intradisc and cytoplasmic space well shielded from the extracellular medium by the plasma membrane.However in fragmented ROS,lacking an envelope,the disc membrane is directly exposed to the suspension medium.Consequently the calcium release capacity strongly depends on the ionic strength of the medium, indicative of a possible control of the calcium release by the surface potential of the disc membrane.After resuspension of fragmented ROS in a low electrolyte medium the calcium release can be almost completely restored.

   In intact ROS,calcium is released with a half-rise time of 3oo ms,whereas after fragmentation the release velocity is greatly enhanced.In sonicated ROS,the half-rise time is 1o ms at 2o°C and closely parallels the kinetics of the Metarhodopsin I/Metarhodopsin II-transition(M I/M II).The M I/M II-reaction shows temperature and pH-dependent biphasic behaviour which can be described by two first order processes.In contrast the calcium release exhibits a monophasic timecourse which matches the slow phase of the M I/M II-transition.For both processes(calcium release and M I/M II-transition in sonicated ROS)an activation energy of about 35 kcal/M has been determined.This suggests that the M I/M II-transition is an important constituent of the calcium release step.

   In contrast,the Arrhenius plot of the light-stimulated calcium signals in intact ROS is nonlinear and can be approximated by two intersecting straight lines reflecting two different processes with activation energies of 11 kcal/M and 35 kcal/M respectively.In the temperature range where intersection occurs,the timecourse of the calcium response exhibits a sigmoid shape with a prominent delay phase typical for two consecutive reactions at temperatures where both reaction steps have similar rate constants. The reaction with the higher activation energy can be attributed to the proper release step by which calcium is released from binding sites during the M I/M II-transition. The lower activation energy probably represents a diffusion controlled reaction which is rate limiting for temperatures above 1o°C.

   It is concluded that the M I/M II-transition is involved in the trigger mechanism of a light-stimulated modification of calcium binding sites in the disc membrane of vertebrate photoreceptors.It is noteworthy that the close relationship between calcium release and the M I/M II-transition does not necessarily imply binding of calcium to the rhodopsin molecule.

   We tend to intepret the slow kinetics of the calcium release in intact ROS as a diffusion controlled step:either a photoactivated rhodopsin molecule diffuses to a locus where it can release calcium or diffusion of calcium ions through the cytoplasm is hindered by multiple rebinding to calcium buffering groups at the surface of the disc membrane(chromatographic effect).Alternatively,calcium release might indicate a light sensitive enzymatic activity in ROS whose rate depends on the structural integrity of its environment.

§ New adress:Lehrstuhl für Biophysik,Univ.Osnabrück,Allg.Verfügungszentrum AVZ,
             Albrechtstraße 28,45oo Osnabrück,Germany

CALCIUM-CONTAINING AND CALCIUM-ACCUMULATING STRUCTURES
IN PHOTORECEPTOR CELLS OF THE LEECH
HIRUDO MEDICINALIS

B. Walz

Abteilung für Vergleichende Neurobiologie (Biologie IV), Universität Ulm
D-7900 Ulm, FRG

A rise and fall in the concentration of intracellular ionized calcium (Ca^{2+} in) is thought to be one factor causing light and dark adaptation respectively in invertebrate photoreceptor cells (1). There is little information about the sources and sinks of Ca^{2+}in. In order to identify intracellular calcium stores in an invertebrate photoreceptor cell, the visual cells of the leech were investigated in a combined unltrastructural, cytochemical and X-ray microanalytical study.

In the visual cells of Hirudo an extensive and highly elaborate three dimensional network of smooth endoplasmic reticulum cisternae is found in very close proximity to the receptive (microvillar) membrane. Structurally similar submicrovillar endoplasmic reticulum cisternae (SMC) are found in most invertebrate visual cells.

Two variants of the potassium pyroantimonate technique (2) were used for subcellular calcium localization. Electron opaque pyroantimonate precipitates were located within mitochondria, associated with SMC, and attached to the inner side of the plasma membrane. The reaction products were identified by energy dispersive X-ray microanalysis (EDX) to contain calcium and antimony. These results show that SMC and mitochondria are major intracellular calcium stores.

To test Ca^{2+}-accumulation into subcellular organelles, saponine-skinned photoreceptor cells were used for an in situ accumulation experiment: Calcium oxalate precipitates (as proved by EDX) in SMC demonstrate that this organelle is able to accumulate Ca^{2+} from a concentration of 2×10^{-5} M, when ATP, Mg^{2+} and oxalate ions are present in the accumulation medium. This result is direct support for the hypothesis that the SMC play an important role in the regulation of intracellular ionized calcium in invertebrate photoreceptor cells. The close proximity between the SMC and the receptive membrane is compatible with a functional relationship between the two structures.

References: 1. Brown, J. E.: Calcium ion, a putative intracellular messenger for
               light adaptation in Limulus ventral photoreceptors. Biophys. Struct.
               Mechanism. 3, 141 - 143 (1977).

            2. Simson, J. A. V. and S. S. Spicer: Selective subcellular localiza-
               tion of cations with variants of the potassium pyroantimonate techni-
               que. J. Histochem. Cytochem. 23, 575 - 598 (1975).

Effects of extracellular Ca^{++} on the response of photoreceptor cells of Hirudo medicinalis

J. Wulf

Universität Ulm

Abteilung für Vergleichende Neurobiologie

Light evoked extracellular responses were recorded from leech visual cells by inserting microelectrodes into the vacuole and were analysed for characteristic descriptive parameters: latency, time-to-peak, half-decay-time from the peak and amplitude. These parameters were measured for relative light intensities varied from threshold through saturation for different adaptation levels maintained by means of constant adaptation flashes given at fixed time intervals. These sets of data, obtained in standard ringer solution, were compared to corresponding data after exchanging to solutions with raised or lowered Ca^{++} content. Omission of external Ca^{++} ($c_{Ca} \approx 2$ μM) leads to the following results:

1. All parameters increase, but the largest effect is on the amplitude and on the half-decay-time.
2. The adaptation level, defined by the half-saturation-intensity in the response-log I-curve seems to be unaffected whereas the slope and the saturation voltage increase.

On the other hand, if extracellular Ca^{++} is raised, the response-log I-curve is somewhat less steep and saturates at lower voltages. Half-saturation intensities, however, remain unchanged. These effects are reversible after exchanging the solution to standard ringer. In standard ringer as well as in Ca^{++}-free ringer the latency, time-to-peak and the half-decay-time vary linearly with the logarithm of the relative light intensity over 1.5 log units.

These results are consistent with a model assuming both intra- and extracellular Ca^{++}-binding sites. Extracellular Ca^{++} apparently modulates the cell's sensitivity within a constant adaptation level, but has little effect on the adaptation level itself, which is suggested to depend on intracellular Ca^{++}.

TIME COURSE AND TEMPERATURE DEPENDENCE OF THE LIGHT INDUCED
ALKALISATION EFFECT IN BOVINE ROD OUTER SEGMENTS

D. Emeis and K.P. Hofmann
Institut für Biophysik und Strahlenbiologie
der Universität Freiburg im Breisgau
Albertstrasse 23, D-7800 Freiburg, W. Germany

Flash photometry of the alkalisation effect in preparations of bovine rod outer segments (ROS) using pH-indicators offers the advantage of time resolution in the msec-range. There are, however, several super-imposed light scattering changes with different dependence on bleaching which therefore cannot easily be eliminated (1). This artifact is avoided in our two-wavelength differential device (2). Thus, we were able to contribute an exact kinetic analysis to the data on this effect known from literature (3,4,5,6). We used the dye bromcresolpurple (λ=590 nm/660 nm) at a pH of 6.

1. Our ROS preparations showed a msec alkalisation and a superimposed slow acidisation not discussed here. The time course of the alkalisation (ΔpH-signal) can be described by a sum of two or three kinetic components.

2. With disk preparations according to Smith et al. (7) the ΔpH-signal is strictly monophasic at all temperatures and coincides with the fastest component in ROS. Neither the Arrhenius plot nor the time course of the ΔpH-signal show the behaviour observed with consecutive reactions. The metarhodopsin signal measured at the same sample after buffering is biphasic as measured previously in ROS (8). The two kinetics are faster and slower resp. than the ΔpH-kinetics at all temperatures.

3. While the amplitude of the metarhodopsin signal increases with temperature according to the metarhodopsin I/II-equilibrium, the ΔpH-signal amplitude remains constant as also observed by Bennett (3).

4. The ΔpH-signal amplitude decays proportionally to the amount of still unbleached rhodopsin as it is observed with metarhodopsin I/II.

As yet, we are not able to explain these results with the known schemes on metarhodopsin I/II and protonation (3,4,9).

1) Hofmann, K.P., et al. (1976) Biophys.Struct.Mechanism 2, 61-77

2) Hofmann, K.P., Emeis, D. (1979) Rev.Sci.Instrum. 50, 2, 249-252

3) Bennett, N. (1978)Biochem.Biophys.Res.Commun. 83, 2, 457-465

4) The papers on this topic cited in (3)

5) Falk, G., Fatt, P. (1966) J. Physiol. 183, 211-224

6) Emrich, H.M. (1971) Z. Naturforsch. 26b, 4, 352-356

7) Smith, H.G., et al. (1975) Exp.Eye Res. 20, 211-217

8) Hoffmann, W., et al. (1978) Biochim.Biophys.Acta 503, 450-461

9) Uhl, R., et al. (1978) Biochemistry 17, 5347-5352

EVIDENCE FOR LIGHT INDUCED LATERAL CONTRACTION

OF THE DISK MEMBRANE BY SMALL BLEACHINGS OF RHODOPSIN

(A MORE DETAILED INTERPRETATION OF THE "P-SIGNAL")

K.P. Hofmann and D. Emeis
Institut für Biophysik und Strahlenbiologie
der Universität Freiburg im Breisgau
Albertstrasse 23, D-7800 Freiburg, W. Germany

Several light induced light scattering changes are observed in the near infrared in suspensions of bovine rod outer segments (ROS) (1). One of them, the socalled P-signal is exhausted within the first few per cent of pigment bleaching. It is caused by a small msec-shrinkage of the ROS (2).

In view of the shrinkage effect seen in X-ray measurements (3), it is suggested that the shrinkage producing the P-signal is an axial effect. On the other hand, we have seen additional slow effects in light scattering (4) which could also correspond to the effect observed in X-ray. We therefore did a reinvestigation of the P-signal as a function of scattering angle. Due to the enhancement in sensitivity of our device, we are now able to measure with low concentrations and to avoid multiple scattering distortion. Thus, a comparison of the signals as a function of scattering angle with differences in particle scattering functions (5) calculated for small changes in the shape of cylindrical particles could be performed. It is seen that the congruence is much better for the assumption of a change in diameter.

In order to obtain further evidence for this result, we did the following experiment: The ROS were oriented by a magnetic field (10 kG) with 90° to the optical axis (6). The P-signal was measured in this way that detection was performed in a slit oriented with 90° to the optical axis but perpendicular respectively parallel to the orientation of the ROS axis. The P-signal amplitude was threefold higher in the former orientation, as was expected in the light of the above results.

Since the outer membrane scattering effects are excluded in the preparation applied, we deduce that the disk membrane undergoes rapid lateral contraction within the first few per cent of pigment bleaching.

1)   Hofmann, K.P., Uhl, R., Hoffmann, W., Kreutz, W. (1976) Biophys. Struct. Mech. 2, 61-77

2)   Uhl, R., Hofmann, K.P., Kreutz, W. (1977) Biochim. Biophys. Acta 469, 113-122

3)   Chabre, M. (1975) Biochim. Biophys. Acta, 382, 322-335

4)   Reichert, J., Hofmann, K.P.: Presented at the Joint Biophysics Meeting Liège, 1977

5)   Malmon, A.G. (1957) Acta Cryst. 10, 639-642

6)   Chalazonitis, N., Chagneux, R., Arvanitaki, A.C.R.Acad. Sci., Paris, 271 D, 130-133 (1970) (cited in (3))

ELECTRICAL PROPERTIES OF THE EXTERNAL LIMITING MEMBRANE IN THE RECEPTOR LAYER OF THE
FROG RETINA

W. Ehrhardt

Zentrum für Physiologie

der Justus Liebig-Universität Giessen

6300 Giessen/Lahn, BRD

Retinal photoreceptors generate currents which can be recorded extracellularly. When
these currents spread along the receptor cells they flow through the external limiting
membrane (e.l.m.), a transversal structure formed by neuroglia cells (Müller cells).
The present study describes the electrical properties of this membrane-like structure
and the resulting interference with the spatial distribution of the receptor currents.
Measurements of currents were made with an array of two or three microelectrodes
which were moved through the receptor layer. Four groups of results were obtained:

i) Large voltage drops were frequently recorded when the electrodes were advanced
   into the e.l.m. of the dark-adapted retina.

ii) Voltage drops between a pair of microelectrodes caused by transretinal current
   pulses increased 3 to 8 times at the depth of the e.l.m. indicating an increase in
   resistance in this region. Locally, up to a 20 fold increase in resistance was observed by means of current injection through the recording electrode. Recordings
   of current spread in the lateral direction suggested that laterally the resistance
   increased by a factor of 3 to 4 at the depth of the e.l.m.

iii) Large photoresponses which accompanied large voltage drops were sometimes observed in regions of high local resistance. The lateral spread of the responses
   was less than 10 μm.

iiii) The resistance of the e.l.m. did not change during light responses.

From these observations one may conclude that the e.l.m. is electrically passive
with respect to the receptor current. However, its behavior is not uniform, i.e.
local areas of high resistance alternate with those of lower resistance.
Given an extracellular resistivity of 100 Ωcm the local resistivity of the e.l.m.
ranges between 300 and 2000 Ωcm. The areas of lower resistance may correspond to
open spaces between the glial cells which comprise the e.l.m.

MECHANISM OF PHOTORESPONSE GENERATION IN ISOLATED FROG ROD OUTER SEGMENTS

W.S. Jagger

Zentrum für Physiologie

der Justus Liebig-Universität Giessen

6300 Giessen, BRD

Photoresponses were recorded from single isolated frog rod outer segments by drawing them into the tip of a tight-fitting glass pipet which served as a recording electrode. The size and behavior of the recorded photovoltage of a fully illuminated outer segment indicate that it is probably due to a proximally flowing external photocurrent causing a potential drop across the medium between the outer segment and the pipet wall.

The shape of a photoresponse from an isolated rod outer segment is very similar to that recorded intracellularly from rods still in the retina. The amplitude of the photoresponse rises steeply with increasing stimulus intensity. The stimulus intensity which elicits responses of half-saturation amplitude is 180 photons absorbed per rod. When responses are recorded over a long period of time, the response amplitude decreases and the duration increases, indicating that the total charge flowing per response remains approximately constant.

When the outer segment is locally stimulated, current originates in the stimulated part and flows into the unstimulated region. If one end of the outer segment is light adapted with a flash, this part remains insensitive to further stimuli, while the other end remains normally responsive. This local adaptation is probably caused by limited diffusion of an internal transmitter along the outer segment.

Possible ionic mechanisms of generation of the photoresponse, which apparently occurs in the absence of a dark current are treated.

CHEMORECEPTION IN THE CRAYFISH ORCONECTES LIMOSUS

H. Hatt and U. Bauer

Physiologisches Institut der Technischen Universität München

Biedersteiner Strasse 29, 8000 München 40

The chelae of the crayfish walking leg bear two morphologically remarkably different structures: large setae combined in clusters which are distributed over the whole carapace, and assymetrical small stout setae lining each biting edge. Afferent neurons originating from both types of setae were found in the connective tissue running just below the carapace. The activity of single fibres was recorded by means of suction electrodes at maximal distance of about 0.1-1 mm (Bauer and Hatt). Mechanosensitive units were mainly restricted to the large setae, whereas chemosensitive units were situated in the stout setae. Recordings from single units allowed to discriminate between the two receptor modalities. By stimulation of mechanosensitive unit, deflecting single large setae or changing the flow velocity of the bath solution, an on-off response was elicited. Application of different chemical stimuli, allowed the differentiation of at least three different types of chemosensitive units, sensitive either to AA, amines or pyridines. The AA receptors responded to 21 different AA and their derivates. They were most sensitive to histidine and alanine with a threshold of less than 10^{-5} mol/l. At concentrations of 10^{-3} mol/l the units responded with a maximal frequency of about 60 imp/s. After a short delay these units showed an immediate increase of the impulse frequency, followed by a complete adaptation within 5-7 s, although the AA was still present. The amine receptors were sensitive to 8 different amines. The threshold concentrations for hydroxylamine and aethylamine were about 10^{-5} mol/l. At concentrations of 10^{-2} mol/l, a maximal frequency of about 30 imp/s was elicited (after 1-2 s delay) which stayed nearly constant as long as the drug was present. The "pyridine" receptors were most sensitive to pyridine, reacting to concentrations of 10^{-6} mol/l - 10^{-2} mol/l. This type of unit was also stimulated by 4-amino-pyridine and nicotinic acid. After chemical stimulation the impulse frequency increased with a delay of 1-2 s and reached a maximum of 50-60 imp/s within about 3 s, followed by a decrease to a tonic level of 10-20 imp/s. This response lasted for more than 10 s after drug application.

Bauer,U. and Hatt,H. (1978) Pflügers Arch. 377, R 50

Section D

RADIATION BIOPHYSICS

A SIMPLE IRRADIATION FACILITY WITH PROTONS

B. Mukherjee

Institut für Strahlungs- und Kernphysik
der TU Berlin - Sekr. KPK 1
Straße des 17. Juni 135 D 1000 Berlin 12

This paper describes a simple and reliable experimental set up for the irradiation of biological objects, such as cell suspensions etc. with protons. The 20 MeV proton beam from the cyclotron was scattered by a nickel foil. The proton charge was collected in a Faraday cup and integrated by a current integrator. A portion of the scattered protons was extracted through a collimator and projected on the object. The energy spectrum and the yield of protons at the exit was measured by a semiconductor detector previously. A lead shield protected the object from the back ground gamma rays. The depth dose distribution and total dose of protons in the soft tissue was calculated according to the methods given in the reference $\underline{/1-2/}$. An average proton dose of $1.07 \times 10^{-3} Gy/\mu C$ for a depth of 3.2 mm in soft tissue was achieved. The organ dose was monitored by the charge collected in the Faraday cup.

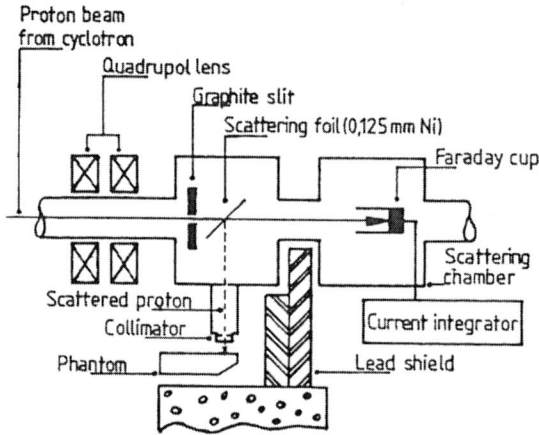

References:
1. Turner, J.E., Zerby, R.L., Woodward, H.A., Wright, W.E., Kinney, W.E. Snyder, W.S., Neufeld, J.: Health Physics Vol. 10, 783 (1964)
2. Kenneth, A.H., Mitchell, J.C., Stewart, J.A. : Radiation Research Vol. 37, 272 (1969)

THE AUTORADIOLYTIC AND THE GAMMA-INDUCED DEMETHYLATION OF THYMINE IN AQUEOUS SOLUTION

O. Merwitz

Institut für Chemie der Kernforschungsanlage Jülich GmbH,
Institut 1: Nuklearchemie, D-5170 Jülich, Postfach 1913

Evidence is presented for the autoradiolytic and for the gamma-
induced demethylation of thymine in aqueous solution. The cleavage
of methyl groups from thymine, in dependence on the concentration
and on the dose rate, was measured quantitatively for the first time
using thymine-(methyl-^{14}C). The demethylation follows a non-linear
dose effect relation.
Radio-gaschromatographic analysis is in hand to determine whether
methane or methanol (or both) are formed during irradiation.
The results are discussed in comparison with the recently investigated demethylation of solid thymine and with respect to their biological significance.

Measurement of irreparable double-strand breaks
in the DNA of eukaryotic cells.

M.Frankenberg-Schwager, D.Frankenberg, D.Blöcher, C.Adamczyk

Gesellschaft für Strahlen-und Umweltforschung mbH, München, Abteilung

Biophysikal.Strahlenforschung, D-6000 Frankfurt, Paul-Ehrlich-Str. 20

The measurement of irreparable double-strand breaks (DSB) in mammalian cells has not yet been possible because the chromosomal DNA molecules are so large. This causes many problems when using the technique of sedimentation of DNA molecules in a neutral sucrose gradient. Yeast cells, however, provide a suitable eukaryotic system in which to measure irreparable DSB since their average relative molecular mass of DNA is about 200 times smaller than that of, for example, human DNA molecules.

Irradiation of diploid yeast cells with 30 MeV electrons results in a linear induction of DNA DSB with dose with a frequency of 0.74×10^{-11} DSB per g/mol per Gy. Most of the DSB are repaired when the cells are kept under non-growth conditions at $30^{\circ}C$ after irradiation. After irradiation with doses which are used for survival studies the repair of DSB is completed within 48 h. The remaining DSB are considered to be irreparable DSB. They are not repaired after the addition of fresh non-growth medium nor during a prolonged incubation period, although the cells per se are still capable of repairing DSB as shown by split dose experiments. The induction of irreparable DSB shows a quadratic function with dose at high dose rate and a linear function of dose at low dose rate.

THE OXYGEN ENHANCEMENT RATIO FOR SINGLE- AND DOUBLE-STRAND BREAKS INDUCED BY TRITIUM INCORPORATED IN DNA OF CULTURED HUMAN KIDNEY CELLS

G.Tisljar-Lentulis, P.Henneberg, L.E.Feinendegen

Institut für Medizin
Kernforschungsanlage Jülich GmbH
5170 Jülich

Asynchronous populations of human kidney T cells were unifilarly labeled with ^{3}H-thymidine during one cell cycle of 28 hours duration. For measuring single-strand breaks (SSBs) the concentrations of the DNA precursor in the nutrient medium of the cells ranged from 5 to 10 µCi/ml leading to specific cell activities between 5,000 and 17,000 disintegrations per day per cell (d/d/c). For double-strand break (DSB) measurements specific cell activities of the order of 10^5 d/d/c were applied. Only high radiation doses were used in the DSB measurements in order to overcome the disturbance caused by the aggregation of high molecular weight DNA in neutral gradients. Decays were accumulated at $1°C$ for time periods ranging from 1 day to 2 weeks in both SSB and DSB experiments. For dose accumulations in the anoxic state petri dishes containing the cells were gassed with a mixture of 95 % pure nitrogen (less than 3 ppm oxygen) and 5 % CO_2. The final partial pressure of oxygen in the gas phase on top of the cell medium was smaller than 5 ppm.

SSBs were measured in alkaline sucrose gradients of pH 12.0. DSBs were measured at pH 9.4. The gradients were centrifuged for 15 to 20 hours at rotational speeds of less than 10,000 rpm. They were calibrated with T_2-phage. For the ratios of known to unknown molecular weights the exponent 0.38 was used. Calculations of DSBs were based on the molecular weight of non-exposed DNA in the neutral gradients (5×10^8 dalton).

In two experiments 1.0 and 1.48 SSBs per ^{3}H-decay per cell were measured in well oxygenated cells. The respective numbers for hypoxic cells were 0.57 and 0.95 SSBs. From these values a mean oxygen enhancement ratio of 1.63 was obtained. The ratio found for DSBs was 3.6 resulting from 0.25 breaks per ^{3}H-decay per cell in oxygenated and 0.07 breaks in hypoxic cells. Data from experiments with ^{60}Co-gamma radiation will be shown for comparison.

REPAIR OF DNA STRAND BREAKS IN CHO CELLS AFTER LOW DOSES

OF X-RAYS

E. DIKOMEY and H. JUNG

Institut of Biophysics and Radiology, University of

Hamburg, Martinistrasse 52, 2000 Hamburg 20

Chinese hamster ovary cells (CHO) were X-irradiated with doses between 200 and 700rad and the number of DNA strand breaks was determined by column chromatography of partly alkali-denaturated and sonified DNA on hydroxylapatite. Using this method, strand breaks after irradiation doses as small as 20 rads were detectable.

The repair of DNA strand breaks at 37°C was determined after doses of 473 and 300 rad by measuring the decrease in the number of strand breaks with time after irradiation. In addition, the number of strand breaks at 37°C was determined immediately and 60 min after irradiation with doses between 200 and 700 rad. The results indicate that the repair of DNA strand breaks can be described by the sum of two exponentiel functions with half-value times of 2.4 min and about 8 hrs, respectively. The amount of strand breaks which are repaired by the slow repair process increases with dose e.g. from 6% to 11% after 300 rad and 473 rad, respectively. When the number of strand breaks one hour after irradiation was compared with the fraction of surviving cells after the same treatment, a good correlation between these two parameters was obtained.

UV EXCISION REPAIR IN LENS EPITHELIAL CELLS DURING AGING
IN VITRO

Roland Zblewski and Hermann Rink
Institut für Strahlenbiologie
Universität Bonn

UV excision repair, usually documented in the form of unscheduled DNA synthesis, represents an important system to prevent radiation-caused disorders in the genetic material (1). When cultured in vitro for some cumulative population doublings diploid mammalian cells either transform spontaneously or degenerate and become senescent (2). This behaviour may possibly be dependent upon the repair capacity. That is why we studied unscheduled DNA synthesis in an in-vitro system during aging and in the transformed stages.

Lens epithelial cells from Sprague Dawley rats, 6 days of age, were cultured serially in M199 supplemented with 20 % of fetal calf serum (Seromed, München) (3). These cells remain diploid (2n=15%). The cells were irradiated (260 nm, NN 15/44, Hanau) with 200, 400 and 600 erg/mm^2 respectively, incubated with 2 uCi/ml ^{3}H-thymidine (24 Ci/mMol, Amersham) for 2, 6 or 24 hs, fixed with Carnoys solution, covered with an Ilford L4 nuclear emulsion, exposed for 14 days in a dark box (4°C), developed by Kodak D19 and stained with a modified Giemsa solution. Grains/nucleus were counted in 100 cells in each assay (n=6). Dosimetry was performed with the Radiometer unit J-2601 (UV Products Inc.). The results show that the UV excision repair capacity remains constant as long as cells in mass cultures remain diploid, but is changed in transformed stages. It is of interest to note, that in the first period after transformation (2n=56 %) unscheduled DNA synthesis decreases, whereas it becomes higher in the final stage (2n=15%).

1) HART,R.W.: Role of DNA Repair in Aging. In: Aging, Carcinogenesis and Radiation Biology. pp 537-556 Ed.: K.C.Smith, 1976

2) HAYFLICK,L.: Fed.Proc.Fed.Amer.Soc Exp.Biol. 34 9-13 (1975)

3) RINK,H., VORNHAGEN,R., KOCH,H.-R.: In Vitro, in press (1979)

RADIATION INDUCED PROTEIN-DNA CROSSLINKS IN EUCARYOTIC CELLS

V.Kasche, U.Frixen, J.Maas, K.Probst, and P.Zipfel
Biology Department, Univ. Bremen
D-28 Bremen 33, F.R.G.

Protein-DNA crosslinks should be formed in radical termination reactions when chromatin in cells is irradiated. Comparatively few studies have, however, been performed to investigate this. The crosslinking of DNA to protein is indicated by indirect evidence. Here a direct method to study this crosslinking is presented.

Tritium was incorporated in DNA of hamster fibroblast cells by incubating the cells with tritiated thymidine in the medium. Using a two-step method that we have developed to isolate protein-free DNA, consisting of a deproteinization with immobilized Proteinase K and an adsorption of DNA to hydroxyapatite, we observed that the absorbance ratio A_{260}/A_{230} decreased with the H-3 content in the DNA. This indicates that protein is crosslinked to DNA due to the decay of this nuclide in DNA. This was confirmed by incubation of the cells in a medium with S-35 labeled methionine and H-3 thymidine. The amount of S-35 associated with the isolated DNA increased with the H-3 content in the DNA.

This method can be used to obtain direct quantitative information on the relative importance, and G-values, of radiation-induced crosslinking of protein to DNA and its repair in cells.

The Effect of Long-Wavelenght UV on BrUra substituted
E. coli: Correlation between Survival, Strand Breaks,
and BrUra-Incorporation.

R. Seidler and W. Köhnlein.
Institut für Strahlenbiologie der Westfälischen Wilhelms-
Universität, Münster, Germany.

If normal E. coli cells are irradiated with 260 nm UV the only photo-
products are thymine dimers ($\hat{TT}$). Using BrUra substituted cells DNA
single strand breaks are induced as well via a debromination process.
Carrier and Setlow (1972) have observed that the relative efficiency
per BrUra of producing single strand breaks is decreasing with increasing BrUra substitution, indicating intramolecular energy transfer.
Using (313) no such energy transfer should take place, and one would
expect a rather simple correlation between loss of biological function
and number of strand breaks in the DNA molecule, since under these
conditions the amount of $\hat{TT}$ produced, which are lethal for the cell,
is negligible.
E. coli C321 (th⁻) was used and cultivated under conditions where 30%
of the thymine is replaced by BrUra and a high colony-forming ability
even after 3 generations in the BrUra medium is maintained. The relative sensitivity of BrUra substituted E. coli C321 as compared to un-
substituted cells has a pronounced maximum at 313 nm UV radiation. The
sensitivity increases with increasing growth time in the BrUra medium
under otherwise constant conditions. Contrary to the expectation the
relative sensitivity per substituted BrUra decrease with increasing
BrUra content in the DNA. The cross-section is apparently not constant.
   The number of strand breaks should be proportional to UV flux and
to BrUra substitution. It was found, however, that the proportionality
holds for the UV flux only. We observed a strong dependence of the
number of strand breaks with growth-time of the cells in BrUra medium.
BrUra substitution was measured using analytical ultracentrifugation
of the DNA in a CsCl gradient. In correlating loss of colony-forming
ability and number of SSB/E. coli SSG we find at a 37% survival 90 -
130 SSB/SSG after 1 generation and 40 - 60 SSB/SSG after 2 generations
growth in BrUra medium. The rather high number of strand breaks per
inactivating event after 1 generation indicates a high repair capacity
which is decreasing after 2 generation growth in BrUra medium.

CORRELATION BETWEEN ENHANCED RADIORESISTANCE OF SPHEROID CELLS AND CELL COUPLING.

H. Dertinger, Kernforschungszentrum Karlsruhe, Institut für Genetik und für
                Toxikologie von Spaltstoffen, Postfach 3640, 7500 Karlsruhe 1

D.F. Hülser, Universität Stuttgart, Biologisches Institut,
                Ulmer Str. 227, 7000 Stuttgart 60

Multicell spheroids have been shown to survive better than monolayers after
irradiation [1,2]. Meanwhile cell lines lacking this "contact-resistance"
have been observed. This gave us the possibility to challenge the hypothesis
that cells in the spheroids could efficiently cooperate through junctions
leading to an enhanced radioresistance. If the hypothesis were correct, only
cells developing "contact-resistance" to radiation should form such junctions. We could indeed demonstrate a correlation between "contact-resistance"
and the existence of gap-junctions mediating ionic coupling between cells.
Several spheroid-forming established cell lines were cultured as monolayers
and spheroids [2]. From the cell survival curves obtained after γ-irradia-
tion the quantity ρ=Spheroid dose/Monolayer dose for 50 % survival was calculated for each cell line. Ionic coupling and electrical membrane resistance were measured using microelectrode technique [3]. ρ = 1 (no "contact-
resistance") was found for the uncoupled cells, whereas coupled cells yielded
values between 2 and 3. Cell cooperation did not correlate with morphology
(epitheloid/fibroblastoid), tumorigenicity, doubling time, DNA content,
chromosome number and growth medium.

[1] Durand, R.E. and Sutherland, R.M., Exptl. Cell Res., 71, 75 (1972)

[2] Dertinger, H. and Lücke-Huhle, C., Int. J. Radiat. Biol., 28, 255 (1975)

[3] Hülser, D.F. and Webb, D.I., Exptl. Cell Res., 80, 210 (1973)

INFLUENCE OF X-IRRADIATION ON THE ROSETTE-FORMING
CAPACITY OF HUMAN LYMPHOCYTES

B.Kessler-Rosbach, H.-J.Meyer-Teschendorf and H.Rink
Institut für Strahlenbiologie
Universität Bonn

T-lymphocytes are known to form spontaneously cellular aggregates,
- so-called rosettes - with heterologous erythrocytes. This phenomenon depends on the membrane receptors of T-lymphocytes. Its alter-
ation may be used to study radiation-induced changes of cellular
membranes.

Human T-lymphocytes were prepared according to Seiler (1) and incubated
with sheep erythrocytes ($4^{o}C$, 1h). Irradiation of lymphocytes (10^{6}/ml)
was performed with irradiation doses of 1, 5, 10 and 15 krad (200 kV,
20 mA, 2 mm Al-filter and a dose rate of 1 krad/min).
After the exposure to 1 krad the rosette-forming capacity increased to
120 % but decreased with higher doses to 50 % of the control value when
measured immediately after irradiation. It is interesting to note that
the rosette formation is strongly dependent on time after irradiation.
It decreases with time (2h) to 32 %, 11 % and 3 % for doses of 5, 10
and 15 krad, respectively. The time dependency shows that the initial
radiation effect is followed by a slow reaction, which results in de-
struction or detachment of the receptors for erythrocytes.
The higher rosette-forming capacity after the exposure to 1 krad may be
compared to the action of neuraminidase, which also results in an
increase of rosette formation. Thus one might conclude that radiation
doses as low as 1 krad have a neuraminidase-like effect on the surface
charge pattern. This idea is supported by measurements of the electro-
phoretic mobility of lymphocytes, which decreases after neuraminidase
treatment, due to a reduction of the negative surface charges.

1)   SEILER,J.R., SEDLACEK,H.H., KANZY,E.J., LANG,W.: Behring Inst.Mitt.
     52, 26-72 (1972)

OBSERVATIONS IN SYNCHRONIZED YEAST CELLS (Sacch.uv.) DURING
X-RAY-INDUCED GIANT CELL FORMATION

C.Baumstark-Khan and H.Rink
Institut für Strahlenbiologie
Universität Bonn

X-Irradiated yeast cells of Saccharomyces uvarum grow under appropriate
conditions (limitation of growth by the added amount of glucose) to a
population of almost pure giant cells (1) with 1-2 cellular divisions
only. The synthesis of DNA continues even after an irradiation dose of
100 krad. Preliminary experiments with sucrose density gradient
centrifugation point to no or only small differences in the molecular
weight of giant cell DNA.

It was of interest to get more knowledge on the DNA content and the DNA
distribution in this type of cells. In order to study this question in
more detail, yeast cells were synchronized by heat treatment according
to Kramhøft and Zeuthen (2) and irradiated afterwards (100 kV, 8 mA,
0.4 mm Al-filter, dose-rate 8.95 krad/min). The growth curves obtained
show a typical synchronized growth pattern. The distribution of DNA in
normal and giant cells was studied by fluorescence staining with
4,6-diamidino-4-phenylindol 2HCl (DAPI). The results show that giant
cells are formed because cellular division remains abortive in that the
DNA cannot be distributed into two distinct portions for the daughter
cell. Similar results could be obtained by treatment of nonirradiated
cells with hydroxyurea.

1) RINK,H.;PARTKE,H.-J.: Radiat. and Environm. Biophys. 12 119-125
    (1975)
2) KRAMHØFT,B.;ZEUTHEN,E.: Methods in Cell Biology Vol.XII pp. 373-380
    Ed.: D.M.Prescott, Academic Press, New York 1975

REPAIR AND RECOVERY IN IRRADIATED YEAST

The effect of protein synthesis inhibitors on liquid holding recovery
in X-irradiated yeast

J. Kiefer, I. Wienhard

Strahlenzentrum der Justus-Liebig-Universität

Leihgesterner Weg 217, D-6300 Lahn-Gießen

Wild-type diploid yeast was exposed to 80 kV X-rays in the presence
and absence of oxygen. Colony forming ability was assayed either by
immediate plating or after a four days holding period in non-nutrient
medium. If cycloheximide was added at a concentration to reduce protein synthesis to less than 10% delayed plating recovery was partially inhibited. The time course shows that this mainly due to a
decrease in the final extent whereas the initial rise in survival is
hardly affected. The recovery inhibition is more pronounced in cells
irradiated under hypoxic, i.e. the oxygen enhancement ratio is significantly lower in liquid held cells.

REPAIR AND RECOVERY IN IRRADIATED YEAST

Mutation induction by ionizing radiation:
influence of oxygen and LET

J. Luggen-Hölscher, S. Rase, J. Kiefer
Strahlenzentrum der Justus-Liebig-Universität
Leihgesterner Weg 217, D-6300 Lahn-Gießen

The haploid wild type strain 211-1a of Saccharomyces cerevisiae was irradiated with 150 kV X-rays under several O_2-concentrations ((O_2)) in the surrounding medium and assayed for survival (colony forming ability) and mutation to canavanine-resistance (colony forming ability on canavanine containing medium).

Six different O_2-concentrations (0%, 1%, 2%, 5%, 7.5%, 21% O_2-saturation) and six doses per O_2-concentration (max.: about 60 Gy) were used. Induced Mutation Frequencies and Oxygen Enhancement Ratios were determined.

The oxygen dependence - both for mutation and survival can be described by the so-called Alper formula

$$OER = \frac{m\,(O_2) + k}{(O_2) + k}$$

For the parameters the following estimates were obtained

$$
\begin{aligned}
\text{Mutation:} \quad & m_M = 1.8 \\
& K_M = 21.2 \ \mu M \ O_2 \\
\text{Colony forming ability:} \quad & m_S = 2.2 \\
& K_S = 17.8 \ \mu M \ O_2
\end{aligned}
$$

There appears to be no statistically significant difference between the values for both processes.

In addition mutation induction by ^{241}Am- -particles was measured. The results show that mutation incidence per dose is increased with this type of radiation as compared to X-rays.

REPAIR AND RECOVERY IN IRRADIATED YEAST

Excision repair of yeast DNA after exposure to UV (254 nm)

K.J. Weber, Helga Waller, J. Kiefer

Strahlenzentrum der Justus-Liebig-Universität

Leihgesterner Weg 217, D-6300 Lahn-Gießen

Excision repair has been shown to be an efficient system for elimina-ting pyrimidine dimers from the genome of UV-irradiated yeast. It is expressed by marked differences in survival between wild type yeast cells and repair deficient mutants. The time course of dimer excision can be directly determined by radioimmunoassay (modified 'Farr-Assay') or indirectly by measuring the recovery of template activity for the transcription of a certain gene (ribosomal RNA).

Antiserum towards UV-irradiated DNA is known to bind highly specifi-cally to pyrimidine dimers. Using this method the relative number of dimers was followed for up to 8 hours after irradiation of wild type cells and repair deficient mutants.

In the transcription assay the ribosomal RNA synthesis has been deter-mined by pulse-chase experiments during an incubation period of one hour. From the resulting dose dependences the recovery of functional integrity of DNA can be estimated in wild type cells compared to cells from excision deficient mutant.

The data obtained with both methods show a pronounced decrease of dimer number on incubation in the wild type, but not in excision negative mutant.

REPAIR AND RECOVERY IN IRRADIATED YEAST

Lethal and mutagenic effects of 254 nm- and 313 nm-radiation on yeast
strains of different repair capabilities

F. Zölzer, J. Kiefer

Strahlenzentrum der Justus-Liebig-Universität

Leihgesterner Weg 217, D-6300 Lahn-Gießen

A haploid wild type strain (211-1a) and two repair deficient mutants
(IG2, excision deficient, and rad 18-1α) of Saccharomyces cerevisiae
were exposed to low fluences of 254 nm- and 313 nm-radiation and tested for survival (colony forming ability) and mutation to canavanine-
resistance (colony forming ability on canavanine containing medium).
The fluences required for reduction of survival to 75% and induction
of $4.5 \cdot 10^{-6}$ mutations per survivor (which was the frequency of
spontaneous mutations) were:

		$F_{75\%}/Jm^{-2}$	$F_{4.5 \cdot 10^{-6}}/Jm^{-2}$
254 nm	211-1a	26	14
	rad 18-1α	2.5	–
	IG2	0.4	0.25
313 nm	211-1a	110	75
	rad 18-1α	65	–
	IG2	25	12

Thus, the wild type strain was about 60 times more resistant than the
excision deficient one to the lethal and mutagenic action of the
shorter wavelength radiation, but only about 5 times more resistant to
the longer radiation. For the rad 18 strain the factors for lethality
were 5 and 2.5 respectively, but there were no mutations at all.
At the same survival level essentially the same mutation frequencies
were observed in the excision deficient strain after irradiation at
either wavelength. 254 nm-radiation induced up to 15 times more mutations in the wild type strain, while 313 nm-radiation induced only
2 times more.
The mutation frequency curves were essentially linear with fluence
below about 10 Jm^{-2} at 254 nm and for all fluences applied at 313 nm
(about 10^{-5} mutations per survivor), but quadratic for the wild type
strain above about 10 Jm^{-2} at 254 nm (about 10^{-4} mutations per survivor). The qualitative behaviour has not changed by liquid holding for
1 - 3 days, but the mutation frequencies were enhanced after 1 day
with no further enhancement after 2 or 3 days.

REPAIR AND RECOVERY IN IRRADIATED YEAST

Survival and mutation induction by heavy ions in yeast cells of different radiosensitivity

F. Schöpfer, S. Rase, E. Schneider, J. Kiefer
Strahlenzentrum der Justus-Liebig-Universität
Leihgesterner Weg 217, D-6300 Lahn-Gießen

G. Kraft
Gesellschaft für Schwerionenforschung (GSI), Darmstadt

Diploid and haploid yeast cells (wild type strains 211 and 211-1a, sensitive mutant rad 52a) were exposed in air to a 208-Pb ion beam to about 0.5 MeV/u specific energy. Strain rad 52a is assumed to be deficient in double strand break repair. Wild type cells were also irradiated with 238-U ions of 7 MeV/u specific energy.

Colony forming ability of the cells was determined as a function of the particle fluence. A dose-effect-curve for the rate of mutations induced by U ions in strain 211-1a was also established, using resistance to canavanine as criterion.

Survival curves are purely exponential in all instances. For both ion beams used, the haploid cells had a larger effective inactivation cross section than the diploid cells. The rad 52a is more sensitive to the heavy ions than the wild type. This means that even for the very high LET_∞ values of the particles used (2.5 to 9.5 MeV/u for the Pb ions, and 13.5 MeV/u for the U ions), the effective inactivation cross section depends less on the geometrical size of sensitive sites than on other properties of the cells, since the amount of nuclear material is the same in the strains 211-1a and rad 52a, and in strain 211 it is about twice as much as in 211-1a.

For the U ions, no delayed plating recovery was observed for the diploid cells, whereas after the Pb ion bombardment the survival of the diploid cells rase by a factor of about 3 after delayed plating recovery.

Mutation induction by U ions was more effective as compared with X rays or with 241-Am alpha particles.

THE KINETIC OF YEAST CELLS WITHIN THE FIRST FOUR GENERATIONS
AFTER IRRADIATION WITH IONIZING PARTICLES

W. Grundler
Gesellschaft für Strahlen- und Umweltforschung mbH
Physikalisch-Technische Abteilung
D-8o42 Neuherberg, FR Germany

For repair of damage caused by irradiation of ionizing particles a cell needs en-
zyms, energy and time. Here, the time required for repair processes is investigated. But not the method of "liquid holding repair" normally applied to yeast and
causing an artifical increasing number of surviving cells is taken into account.
Here the segmentation of single cells within the first four generations after irradiation are observed and the time behaviour from budding to budding is registrated
in dependence of energy dose and ionization density. These time intervalls are
self controlled.

The lagtime (control ~ 3 h), i.e. the mean time interval between irradiation of the
stationary early G_1-cells and the first "segmentation" (the appearance of the bud
just before the S-phase)

.   is extended by the energy dose by a factor of 1,45 $\pm$ o,15. This maximum of ex-
    tention is independent of the used radiations, but achieved with a different
    dose; that means for the physiological processes of the G_1-phase the deposited
    energy dose is the important parameter and not the ionization density.

.   The lagtime is equal for dead and surviving cells. Ionizing radiation affects
    the lagtime generally in an other way than it will react on survival.

After the first segmentation the time interval to the second one is elongated too.
One group of irradiated cells divides approximately after one generation time of the
undisturbed population. The remaining double cells mostly surviving (for a dose of
o,5 kGy of fast electrons all the 7o % survivors, for instance) divide much later
($\sim$1o h). But a cell of a double group cannot start the second cycle at any time.
It has to wait at least one cell cycle period or a multiple of it.When for yeast
cells this critical segmentation is performed, establishing a colony of three or
four cells the third division occurs approximately after a normal generation time
($\sim$1,3 h).

It is likely to occur repair of potential lethal damage within the periodical time
interval of the double cells before the third division. It has been shown /1/ that
exponential growing cells have finished for instance the repair of DNA double strand
breaks within a time interval of 6 h after irradiation. This time corresponds with
the period ($\sim$1o h) of the irradiated stationary cells observed as single cells.

/1/ Resnick, M.A.: Molecular mechanisms for repair of DNA. Part B, 459-556 (1975)

THE POSSIBLE ROLE OF C-AMP IN THE IRRADIATION INDUCED CONTRACTION OF

NERVE MUSCLE PREPARATION

L. Schachinger, H. Klöter, M. Michailov and Ch. Schippel
Abt. f. Strahlenbiologie, Gesellschaft für Strahlen- und Umweltforschung, 8042 Neuherberg, Fed.Rep. Germany

c-AMP acts as a key-substance in transforming the stimulation by trans-
mittors via cell receptors into mechanical and electrical cell activities. ß-adrenergic receptors activate the adenylate cyclase, leading
to an increase of the c-AMP concentration and to a relaxation of the
nerve-muscle preparation. By addition of dibutyryl-c-AMP to preparations of frog lung and rat aorta the radiation induced contraction
could be reversed. This ability of c-AMP is lost by X-irradiation in
vitro, probably due to changes of the molecular structure. The physiological activity decreases far more than it would be expected from
measurements of the absorbance at 260 nm which is attributed to the
intact purine ring. Thin-layer chromatograms show substances of lower
R_f-values, produced during radiolysis by addition of OH radicals and
H-atoms. Reflexionspectra and difference spectra (irradiated minus
unirradiated c-AMP) point into the same direction. The effect of the
radiation products is compared with the effect of a shift in the
c-AMP/AMP equilibrium within the cell. Experiments with frog lung preparations showed that the addition of 10 % AMP to Dib-c-AMP (total
concentration 10^{-4} M) converted the relaxation of approximately 30% into
a contraction of ca 11 %.
At a percentage of 20
(AMP in c-AMP), about
20% contraction was
reached.

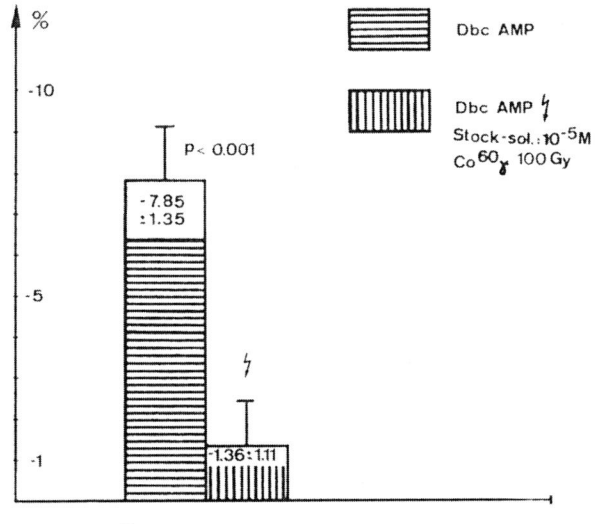

RELAXING EFFECT OF NON-IRRADIATED AND IRRADIATED ϟ
DIBUTYRYL CYCLIC (DBC)- A M P-SOLUTIONS

ON THE SMOOTH MUSCLE TONE OF FROG LUNGS IN %
TO INITIAL LENGTH(L) OF PREPARATIONS.

%

Dbc AMP

Dbc AMP ϟ
Stock-sol.:10^{-5}M
Co^{60}ϟ 100 Gy

-10

P< 0.001

-7.85
:1.35

-5

-1

-1.36:1.11

L· 13.7 :1.3 mm   n·10

IRRADIATION AND GASTRIC EMPTYING

A. Kindt, E.L. Sattler, A. Schraub

Strahlenzentrum der Justus-Liebig-Universität

Leihgesterner Weg 217

D-6300 Giessen

The radiation prodromal effects like anorexia, fatigue and emesis may play a non negligible role in civil nuclear accidents as well as in war. Our findings of the not reduced effect of protracted irradiation as well as other facts indicate, that the radiation prodrom is more or less independent from the effect on cell multiplication.

Therefore we are looking for the physiological cause for the onset of anorexia in rats. The anorexia is combined by a delayed emptying of the stomach. The dose dependence of the stomach emptying was observed by taking photographs using barium sulfate enriched fodder. We than tested several substances, which may be the cause of anorexia or counteract it. The effect of the investigated substances is shown in table 1. For this investigation we used a radiation dose of 400 R. Following drugs were examined: AET (a), Serotonin (b_{1-3}), Clanobutin (c), Cimetidin (d), Dexchlorpheniramin-Maleat (e).

Table 1:

Control physiol. NaCl	a	b_1	b_2	b_3	c	d	e	group
1 ml	150	25	32.5	40	20	100	20	drug dose (mg/kg)
-	++	-	++	++	-	-	-	shame irradiated
++	++	++	++	++	++	++	++	irradiated (400 R)

++ gastric emptying delayed

- normally (control)

The Effect of Neutron- and X-irradiation on the Development of Preimplanted Mouse Embryos

C. Streffer, M. Molls, N. Zamboglou and D. van Beuningen, Institut
für Med. Strahlenphysik und Strahlenbiologie, Universitätsklinikum
Essen, Hufelandstr. 55, 4300 Essen 1

Preimplanted mouse embryos were cultivated in vitro from the 2-cell
stage until hatching. The development of the embryos and the cell
proliferation during this development were investigated by techniques as cytofluorometry, labelling index etc. which were described
earlier (1). The embryos were irradiated with X-rays (12.5-200 R)
and with fast neutrons from a cyclotron (12.5-100 Rad).

After X-irradiation the dose effect curves for the various parameters
demonstrate an appreciable capacity of recovery processes. This is
not the case after neutron irradiation. Therefore the RBE values
depend very much on the radiation dose, they are highest (7-8)
for the cell proliferation at low radiation doses. The comparison
of the cytofluorometric data with labelling by ^{3}H-thymidine for the
same cell nuclei demonstrates that some cells in S-phase are un-
labelled and on the other hand some cells in G_1-phase are labelled.
The possible mechanism of these findings will be discussed.

Further the formation of micronuclei as an expression of cytogenetic
damage is evaluated for both radiation qualities. The number of
micronuclei increases dramatically after irradiation with low neutron
doses.

1) C. Streffer, D. van Beuningen, M. Molls, N. Zamboglou und
   S. Schulz: Kinetics of cell proliferation in the preimplanted
   mouse embryo in vivo and in vitro. Cell and Tissue Kinetics
   (in press).

IMMEDIATE EFFECTS OF IONIZING RADIATION IN RABBITS

Werner, E., Pohlit, W. and Schalk, K. P.

Institut für Biophysik der J. W. Goethe Universität und Ges. für Strahlen- und Umweltforschung, 6000 Frankfurt am Main

Within the first minute of an irradiation of sufficient high dose rate of ionizing radiation reproducible effects on physiological parameters and serum electrolytes are observed. It was the aim of our studies to find out the underlying mechanisms of these effects.

Male rabbits (kleine Gelbsilber, body weight 1 800 g) are kept under anesthesia. Blood pressure is measured by arterial catheterisation and pulse rate derived from the blood pressure curve. Respiration rate is deduced from changes of air pressure in a tube, which is localized between the thorax of the animal and a lucide irradiation cage. Also electrolyte concentrations (K^+, Na^+, HCO_3^- and Cl^-) in blood serum are measured continuously by a modified AutoAnalyzer. Animals are irradiated with 30 MeV electrons from a betatron. About 50 seconds after beginning of irradiation a reversible decrease of mean blood pressure of about 25 % and an increase of pulse rate are observed. About 6 minutes after start of irradiation blood pressure reaches the original level. At the same time respiration rate increases. Sodium and chlorid remain nearly constant. A dose of about 1 krad causes an increase of K^+ of about 60 % and diminishes distinctly HCO_3^- concentration. All extent of these changes are dependant on dose but also on dose rate. The minimal dose required to detect immediate reactions is about 500 rad and a minimal dose rate of about 200 rad/min. These effects occur only at the beginning of the first irradiation if a series of doses are given. If maximum effects are produced by the 1th dose even after a recovery time of 2 hours no effects are seen at the beginning of subsequent repeated irradiations.

Therefore it seems that direct damage of cell membranes as an effect of irradiation can be excluded. We suggest that by irradiation mediator substances are liberated which act on small blood vessels and on cell membranes. In the small vessels permeability may increase with a consequent diminished circulating blood volume and a fall in blood pressure. By way of compensation heart frequency will accelerate. Cell membranes may also be effected. The permeability to K^+ and H^+ may increase. As a result the concentration of serum potassium is elevated and the blood pH decreases. The latter effect is a strong central stimulator for breathing. For that reason hyperventilation occurs which reduces blood bicarbonate concentration.

INTERACTION OF 40,5°C- HYPERTHERMIA WITH 200 kV X-RAYS AT TWO DIFFERENT DOSE-RATES.

U. Schrader-Reichhardt, B. Markus, B. Papenberg; Abt. Strahlenphysik und -biol. der Radiol.-Univ.-Klinik, D 34 Göttingen, v.-Sieboldstr.3

Hyperthermia combined with ionizing radiation results in a strong sensibilization of the radiation effect. The mechanisms responsible for this thermal enhancement are not known exactly. It is often supposed that the interference of hyperthermia with repair of radiation induced damage may be an important factor for thermal enhancement.

Our experiments with CHO cells at two dose-rates have shown:

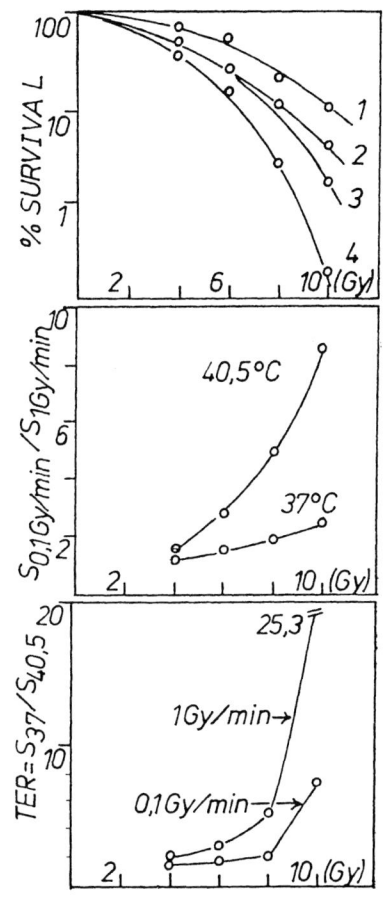

1. The synergistic interaction of hyperthermia with radiation works at relatively low temperatures (40,5°C).
   Curve 1 : 37 °C   0,1 Gy/min
   Curve 2 : 37 °C   1   Gy/min
   Curve 3 : 40,5°C 0,1 Gy/min
   Curve 4 : 40,5°C 1   Gy/min

2. Reducing the dose-rate by a factor of 10 results in higher survival for 37 and 40,5°C. The increase in survival is greater for 40,5 than for 37°C.

3. Thermal enhancement ratios (TER) are dependent on dose and on dose-rate.

Summary

Although thermal enhancement of cell killing is well documented, the sparing effect of low dose-rate is greater for 40,5 than for 37°C: a result, difficult to explain only in terms of repair inhibition of radiation induced damage. Experiments using higher temperatures and higher dose-rates are in progress.

LATE EFFECTS FROM LOW LET RADIATION IN THE LOW DOSE RANGE - THE EXPERIENCE OF HIROSHIMA AND NAGASAKI

I. Schmitz-Feuerhake und E. Muschol
Universität Bremen, Fachbereich Physik
Kufsteiner Str., 2800 Bremen

The JNIH-ABCC Life Span Study of Hiroshima and Nagasaki contains a sample of about 80 000 persons in the low dose range. They were classified by the distance from the hypocenter or by the time of entering after the bombing to be in the groups 0 rad, 1-9 rad, and "Not in city (NIC)". All of them must have been affected by fallout radiation. The dose of fallout burden was estimated to be about 10 rad. Significant effects in these groups have been found for all cancers related to radiation, if the Japanese national rate is taken as a control. A risk factor of about $600 \cdot 10^{-6} \mathrm{rad}^{-1}$ is obtained for all malignant diseases, which is six fold the value given by the UNSCEAR report of 1977 and the ICRP publication No.26 as an upper limit for chronical low LET and low level radiation. The possible errors of the estimates for the special groups of cancers are discussed.

A NON-LINEAR MODEL OF THE GLOBAL CARBON CYCLE
FOR THE ASSESSMENT OF LONG-TERM EFFECTS OF CO_2
AND C-14 DUE TO VARIOUS ENERGY SCENARIOS

M. Matthies, H.G. Paretzke, W. Jacobi

Gesellschaft für Strahlen- und Umweltforschung mbH
Institut für Strahlenschutz
D-8042 Neuherberg

The global cycling of carbon dioxide from the combustion of fossil fuels and of
radiocarbon released from nuclear power facilities has been simulated using a seven-
box-model. The model is built up by two boxes for the atmosphere (stratosphere, troposphere), three boxes for the ocean (mixed surface layer, deep sea and sediments),
and two boxes for the biosphere (short and long-lived biota) with non-linear tropo-
sphere-biota and troposphere-ocean surface layer exchange rates and linear fluxes
between the other reservoirs. Regarding the man-made reduction of the biomass, two
different models are used: (a) with no deforestation, and (b) with an logistic growing deforestation function. The biota growth factor, the exchange of the atmospheric
CO_2 with the ocean, and the preindustrial atmospheric CO_2 content were fitted using
the records of the atmospheric CO_2 concentration in Mauna Loa, the Suess-effect until
1954, and the response to the C-14 from nuclear weapon tests.

The three scenarios considered are (1) annual energy growth rates between 1% and 4%,
no nuclear power; (2) a high and (3) a low increase of energy consumption, each with
low and high substitution of fossil fuels by nuclear power. In addition, two assumptions concerning the decontamination of C-14 in the nuclear power plant effluents
are made, namely a pessimistic one in which all produced C-14 are released, and an
optimistic one with an decontamination factor of 4. Assuming logistic source functions for the increase of fossil fuel combustion and an exponential growth of nuclear
power until the year 2000, the CO_2 concentration of the troposphere reaches the 2 - 5
fold of the preindustrial level around 2100. Simultaneously, the specific C-14 activity of the atmosphere is decreased. Up to 2200 the specific activity is lower than
the preindustrial level. Calculations above this time interval are highly uncertain
and depend primarly on the assumptions made for the exchange with the biosphere.
However, simple estimations show that the maximum C-14 level does not exceed the
1.5 fold of the preindustrial level. For the most pessimistic scenario (including
atomic bomb C-14) the collective dose commitment in 2100 is estimated to 151 man-rem
per Ci released to the atmosphere.

140

AUTHOR INDEX

Name, Initial   Name, Initial

Abrahamson, E.W., 79, 80   Dikomey, E., 120
Ackermann, Th., 11   Dorst, H.-J., 62
Adam, G., 92, 93   Drikos, G., 3
Adamczyk, C., 118   Dudel, J., 85, 86
Akutsu, H., 30   Edelmann, L., 102
Albrecht, O., 23   Ehmann, G., 44
Appleby, C., 15   Ehrhardt, W., 112
Bahr, A., 20   Emeis, D., 110, 111
Bamberg, E., 53, 60, 61   Fahr, A., 61
Bauer, P.J., 106   Feinendegen, L.E., 119
Bauer, U., 114   Felle, H., 75
Baumstark-Khan, C., 126   Finger, W., 85, 86
Beck, H.-P., 95   Frankenberg, D., 118
Benz, R., 42, 43, 44, 62, 63, 64,   Frankenberg-Schwager, M., 118
65, 66   Franklin, R.M., 30
Bernhardt, J., 82   Frehland, E., 48, 50, 51
van Beuningen, D., 135   Frings, D., 105
Blöcher, D., 118   Frixen, U., 122
Blume, A., 26   Fromherz, P., 40
Boheim, G., 54, 55, 56   Funk, J., 77
Borys, T., 79, 80   Gally, H.U., 32
Brock, W., 46   Gersonde, K., 15
Brückner, H., 55   Gödde, J., 89
Bruns, M., 104   Gohl, W., 16
Büldt, G., 31, 32   Gräber, P., 71, 73
Burckhardt, G., 90   Gramlich, V., 11
Carius, W., 87   Greulich, W., 97, 98
Cherry, R.J., 34   Gruenewald, B., 26, 36
Christahl, M., 15   Gruler, H., 23
Claßen-Linke, I., 104   Grundler, W., 132
Conti, F., 83, 84   Grunwald, P., 7, 8
Craubner, H., 57   Gunsser, W., 7, 8
Deitmer, J.W., 88   Haase, A., 40
Dencher, N.A., 34, 35   Hanke, W., 54, 55
Dertinger, H., 124   Hatt, H., 114
Di Fiore, D., 70   Heinen, H.J., 105

Name, Initial

Name, Initial

Henneberg, P., 119

Hermetter, A., 41, 42

Hermann, U., 29

Hertel, C., 76

Heyn, M.P., 33, 34, 35

Hodapp, N., 69

Hofmann, K.P., 110, 111

Hülser, D.F., 124

Huber, H.-L., 74

Jacobi, W., 139

Jagger, W.S., 113

Jähnig, F., 38

Janko, K., 53

Jauch, P., 42

Jordan, P.C., 46

Jung, G., Münster, 12, 13

Jung, G., Tübingen, 55

Jung, H., 120

Junge, W., 78, 107

Junges, R., 47

Jürgens, E., 24

Karsten, W., 21

Kasche, V., 122

Kaupp, U.B., 78, 107

Kessler-Rosbach, B. 125

Kiefer, J., 127, 128, 129, 130, 131

Kindt, A., 134

Klöter, H., 133

Klump, H., 11

Knäble, Th., 11

Knoll, W., 25

Kohl, K.-D., 2

Kohler, H.-H., 52

Köhnlein, W., 12, 13, 14, 123

Kolb, H.-A., 47, 48, 49

Kreutz, W., 5, 6, 69, 77

Lawaczek, R., 39

Lewis, R.S., 12, 13

Lohmann, W., 16, 17, 18, 19, 20, 27, 97, 98

Luggen-Hölscher, J., 128

Maas, J., 122

Maaß, G., 29

Mäntele, W., 5, 6

Markus, B., 137

Marmé, D., 76

Matthies, M., 139

Meraldi, J.-P., 37

Merwitz, O., 117

Meyer-Teschendorf, H.-J., 125

Michailov, M., 133

Molls, M., 135

Mukherjee, B., 116

Müller, A., 21

Murer, H., 90

Muschol, E., 138

Nakae, T., 63

Neher, E., 83

Neubacher, H., 17, 18, 19

Neumann, E., 9, 82

Paltauf, F., 41, 42

Papenberg, B., 137

Pappert, G., 62

Paretzke, H.G., 139

Pasternak, J., 79

Penka, V., 20

de Peyer, J.E., 88

Pilwat, G., 67, 99

Platt, D., 96

Pleyer-Weber, A., 27

Pohlit, W., 136

Probst, K., 122

Raap, A., 15

Rabl, C.-R., 10

Rase, S., 128, 131

Rempel-Rossleben, U., 91

142

Name, Initial

Name, Initial

Renger, G., 71
Rink, H., 121, 125, 126
Rögner, M., 73
Rumberg, B., 74
Rüppel, H., 3
Sackmann, E., 23, 24, 25
Sapper, H., 16, 27
Satake, H., 30
Sattler, E.L., 134
Seelig, A., 32
Seelig, J., 30, 32, 37
Seher, J.P., 92
Seidler, R., 123
Seliger, H., 11
Semple, N., 79
Shepherd, J.C.W., 31
Siebert, F., 5, 6
Smith, Jr., H.G., 106
Spalink, J.-D., 81
Sperling, W., 2, 3
Schachinger,L., 133
Schalk, K.P., 136
Schallreuter, D., 9
Schatz, G.H., 73
Schindler, H., 58, 59
Schindler, R., 43
Schippel, Ch., 133
Schlitter, J., 37
Schlodder, E., 72, 73
Schmid, E.D., 11
Schmidt, K., 66
Schmitz-Feuerhake, I., 138
Schnabl, H., 67, 68
Schneider, A., 60
Schneider, E., 131
Schneider, F.W., 4
Schnetkamp, P.P.M., 78, 107
Schober, E., 7
Schöpfer, F., 131

Schrader-Reichhardt, U., 137
Schraub, A., 134
Schreiber, J., 97, 98
Schröder, W., 104, 105
Schubert, D., 62
Standke, K.-H.C., 100
Stankowski, S., 36
Stark, G., 45, 46
Steiner, U., 93
Stephan, W., 51
Stettmeier, H., 86
Stieve, H., 104
Streffer, C., 135
Strobelt, W., 27, 96
Stühmer, W., 84
Stuhrmann, H.B., 25
Stulz, J. 11
Thurm, U., 89
Tiemann, R., 70, 71
Tisljar-Lentulis, G., 119
Tobüren, D., 94
Tretzel, J., 4
Tümmler, B., 29
Uhl, R., 79, 80
Oberschär, S., 56
Vienken, J., 68, 99
Vogel, H., 28
Vögtle, F., 29
Vömel, Th., 96
Wallbott, W., 17
Waller, H., 129
Walter, D., 69
Walz, B., 108
Watanabe, F., 26
Weber, K.J., 129
Werner, E., 136
Weisenseel, M.H., 101
Welte, W., 69, 77
Wienhard, I., 127

Name, Initial	Name, Initial
Witt, H.T., 70, 72	Zaplatynski, P., 18, 19
Woermann, D., 49	Zblewski, R., 121
Wood, P.G., 91	Ziegler, H., 68
Wulf, J., 109	Zimmermann, F., 14
Wutschel, I., 77	Zimmermann, U., 64, 65, 67, 68, 99
Zaccai, G., 32	Zipfel, P., 122
Zamboglou, N., 135	Zölzer, F., 130

Biophysics
of Structure
and Mechanism

ISSN 0340-1057   Title No. 249

Editors: L. Brand, H. Eisenberg, F. Sauer, G. Schwarz, H. Stieve (Managing Editor)

This journal of physical biology publishes results of experimental studies and theoretical studies on the following range of subjects:

Molecular structure, structural change and its biological function. This should be the main field of the journal. It includes X-ray structure analysis, spectroscopic studies, such as NMR, CD, ORD, IR, VIS, UV, and ESR, kinetics of reactions, interactions of molecules, etc.

Transport phenomena and thermodynamics of irreversible processes applied to biological phenomena. Since there exists a great number of journals which deal with the field of membranes, this journal will restrict itself to membrane biophysics in which emphasis is placed on the more physical approach to the membrane.

Primary reactions in photosynthesis and sensory transduction.

Springer-Verlag
Berlin
Heidelberg
New York

Subscription information and sample copies upon request

Radiation and Environmental Biophysics

ISSN 0301-634X

Title No. 41

Managing Editor: H. Muth, Homburg, Saar

Editors: V. P. Bond, Upton, NY; F. Dunn, Urbana, IL; H. Fritz-Niggl, Zürich; A. S. Garay, College Station, TX; H. Glubrecht, Hannover; A. W. Guy, Seattle, WA; U. Hagen, Neuherberg; A. Kellerer, Würzburg; K. R. Knoerr, Durham, NC; H. Lieth, Osnabrück; S. M. Michaelson, Rochester, NY; P. Oftedal, Blindern; H. Pauly, Erlangen; H. J. Schaefer, Pensacola, FL; H. P. Schwan, Philadelphia, PA; C. A. Tobias, Berkeley, CA; K. Wagener, Jülich

The journal is devoted to fundamentals as well as to applications. Its range of interest includes:

1. Biophysics of ionizing radiations (including radiation chemistry, radiation cytology, radiation genetics; incorporation of radionuclides; biophysics as the basis of medical radiology, nuclear medicine and radiation protection).

2. Biophysics of nonionizing radiations: ultraviolet, visible and infrared light (including laser light), microwaves, radio waves, sound and ultrasound.

3. Biological effects of such physical factors as temperature, pressure, gravitational forces, electricity and magnetism.

4. Biophysical aspects of environmental and space research, i.e. physical parameters used in the description of ecosystems, mathematical treatments of models of the environment, mechanism and kinetics of flows of matter and energy in the biosphere.

The treatment of these themes may include both theoretical-mathematical and experimental material, and can embrace complex radiobiological phenomena as well as health physics and environmental protection.

Special emphasis is given to fundamental questions. Papers on medical physics or biomedical engineering will be accepted if they contribute to the understanding of biophysical mechanisms.

The journal accepts also important papers which are not concerned directly with effects of radiation or environmental problems but with the scientific basis for their understanding, e.g. biophysics of genetic processes and mutations, mathematical models of biological or environmental systems and the mathematical treatment of cell kinetics.

Springer-Verlag
Berlin
Heidelberg
New York

Subscription information and sample copies upon request.